美國油匠在台灣

1877-78年苗栗出礦坑採油紀行

American Oil Technicians in Formosa:

A Record of Drilling Oil at Chhut-hong-khinn in Miaoli, 1877-78.

陳政三 著

目 次

再回出磺坑：永懷陋巷歲月

2005年10月14日，拿到剛剛出爐的第一版《出磺坑鑽油日記》（第一版原書名，歷史智庫出版），在內頁寫下「挖出被掩埋的史實，就如沙漠遇甘泉，也如簡時、絡克鑽打到油脈、流出第一滴油的心情……」

1878年（光緒四年）陽曆8月3日，油匠終於鑽打到油層，雖然流出的鹽水遠多於石油，絡克仍特別在當天日記標上「1878」，以紀念全大清國、全亞洲，也是台灣地區有史以來用機器鑽出「第一滴油」的大日子。不過種種工作上的不順，加上異鄉生活不容易，疾病頻傳，9月3日與清國官方一年合約期滿之後，該月15日簡時、22日絡克相繼生病。合約規定洋技師生病須自行就醫，意即「萬一病死，官方不負責任」，更加深他們不再續約的念頭。27日病懨懨的絡克搭轎永別出磺坑油井，29日抵達大稻埕。簡時多待在油井10天，稍後也是抱病北上。就此結束此段台美不愉快的合作經驗。

中影國片《源》（1979），描寫客家移民在苗栗山區出磺坑開鑿石油的故事，裡面兩位外國人飾演跑龍套式的美國油匠角色即是簡時與絡克；次年發行了VCD片，內容大抵根據地方方志與傳說，與史實相去甚遠。近期客家電視台重拍《源》連續劇（2009），大部份情節仍參考張毅電影劇本原著，原製作人余崇吉曾和筆者聯繫，說要採用本書部分情節，談妥之後不久，筆者因故遷回中部，他又離職，因此，後來客家電視台就未採用本書情節，有點可惜。

之前，台灣歷史相關中英文書籍，對於光緒三年～四年（1877～1878）那段開鑿「大清」中國，也是台灣有史以來首座油井的記載，不是完全未提，就是一筆帶過，更未提到兩位老美技師的中英文姓名。

大清〈1877年淡水海關英文年報〉上載：「在唐景星（Tong King-sing, 唐廷樞）主辦下，聘請兩位美國人，添購機器，準備在後壠（苗栗縣公館鄉出礦坑）開鑿油井」；〈1878年淡水海關年報〉記載甚詳，將整個載運機器、鑽油過程、油井深度、產油數量、遭遇的困難，以及美國技師離開淡水返美的日期寫得十分詳細，但就是少了油匠的名字。翻索達飛聲（James Davidson，在台灣使用「德衛生」）巨著*The Island of Formosa, Past and Present*（1903），裡面只寫「招募兩名美國技師來台鑽油」。《淡水廳志》、《苗栗縣志》也只簡單交代初期民間與洋人在苗栗山區爭奪採油權的糾紛；大清相關檔案可能在鑽油計畫失敗後全銷毀了。

筆者有次翻閱Harold M. Otness著，中研院台史所出版的*One Thousand Westerners in Taiwan, to 1945*，意外發現了「Robert D. Locke, American oil technician」的記載。根據這個線索查到Karns，發現中研院近代史研究所已經購得油匠絡克（Robert Locke）的原始日記及相關文件；不過追查的結果，該所館藏資料居然憑空消失！

繼續查詢Otness書中提到Sampson Hsing-chang Kuo寫的一篇美國喬治城大學博士論文：*Drilling Oil in Taiwan: A Case Study of Two American Technicians' Contribution to Modernization in Late Nineteen-Century China*（1981）；終於在Mr. Kuo（郭先生？）的論文末，發現附有Paul H. Giddens整理過的絡克《鑽油日記》，日期從1877年9月4日離開家鄉，到隔年12月15日返抵美國爲止。1940年，研究美國石油發展史長達半世紀的Giddens曾經採訪絡克，後者當時已經高齡90，G君取得絡克的日記原件，以及美國技師與大清台灣府當局的簽約文件、支付薪水證明、匯票水單等原始文件，最後把上述文件都賣給了台灣的中研院（Kuo, p. 463）。

另外，本書初版付梓後，又在《中央研究院近代史研究所大事紀要》（台北：中研院近史所，1985，頁104～105）民國六十八年五月條，發現上載：「美國前翰林大學（Hamline University）校長吉丹斯博士（Dr. Paul H. Giddens）同意將其所存一八八七（按一八七七至一八七八）年美籍技師開採后龍油井原始資料多件，讓售於本所。」不過，正如前述，「通通不見了！」

初版甫上市，照例訪遍朋友，大力推銷此書，遭過「毒手」的親友，都見識過我賣書的狠勁，面對簽過名的書，加上「萬一死後有名，這本書可值錢喔！」名言，常啼笑皆非，乖乖掏錢。廖榮隆、林明福是「受害」最深者。當然，也有陌生人捧場。如此不要臉推銷，用意在答謝願意冒險的出版社。過程中有人建議，「爲何不同時收錄原文？中英對照除保存原始產業文化資產史料，也可讓讀者更加能夠掌握原意。」好建議。但可得有再版機會啊。

撇開看書、買書風氣不佳的大環境，出版社原就生存不易，類似書籍想再版的可能性不大。幸虧，筆者之前的勞作，都有再版的情形。本想「以時間換取空間」，長銷書總有「醜媳婦熬成婆」的機會，等哪天賣得差不多，有機會再版時，再提中英對照修訂版建議。可惜，2009年12月，《歷史月刊》發行最後一期（第263期），宣佈停刊，歷

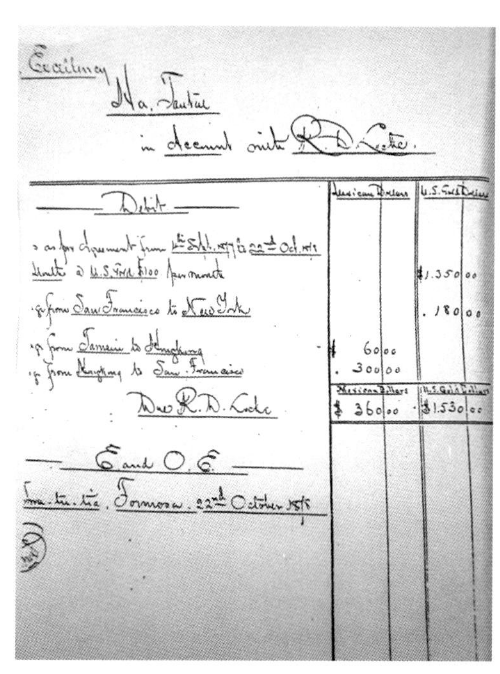
in Account with R. D. Locke.

Debit

from San Francisco to New York

Mexican Dollars	U.S. Gold Dollars
	$1.350 00
	. 180 00
$ 60 00	
. 300 00	
$ 360 00	$1.530 00

R. D. Locke

E and O. E.

Formosa. 22nd October

▲夏獻綸發給絡克之薪水、旅費明細表（Sampson Kuo書，陳政三翻拍）

史智庫出版社稍後也停止營運，令人唏噓。筆者從該刊讀者到作者的過程，宛如夢境。感謝時任發行人東年大哥的青睞、提攜，終生銘感。

「中英對照」之外，修訂版係反映筆者、同好與讀者對內容的看法，譬如，中研院台史所老友翁佳音對於筆者將1878年10月5日「tea house」解讀為「茶舖」認為有待商榷。經再三斟酌、推敲，於是將之更改為「茶樓」，並加註：「第一版一刷筆者誤解讀為『唐景星在北台經營茶葉生意』；『tea house』似應為接待人客的茶樓」。佳音兄根據荷蘭史料、金門地名與移民台灣歷史途徑，指出「後壠」地名應該是來自金門（東邊金湖鎮）的「甌壠」。其他尚有筆者對人名、地名、河流等名稱的更正，加註出處、頁數等，用意在讓全書更正確、更忠於史實，也讓讀者可以找到註釋的原始出處。最重要的是，把第一處開鑿的出磺坑油井修正為位於後龍溪南畔。初版解讀油井在「溪北」，本書更正為「溪南」之因，是由於完整的《馬偕日記》之刊行，記載了油村名「溪洲」，筆者乃重新思考、反覆推敲之後，所做的修正。另外，書末加入索引，不但讓讀者，也讓筆者能夠較容易翻索想找的資料。

2009年初，筆者再度「誤入歧途」，被長期關照再三的老長官王主任壽來「誘拐」到位於台中的「文建會文化資產總管理處籌備處」（2012年5月20日起，更名為「文化資產管理局」）服務，一方面遷回彰化老家，陪伴高堂老母，嘗試痛改前非，略盡為子之道；另方面暫脫離居陋巷不改其志，不改其憂，也不改其樂的生活。業務範圍恰與出磺坑有關，重回該地，突興起「念天地之悠悠」感慨。

書中老美油匠從烏坵島帶來台灣的狗狗「薯條」（1877年11月22日、12月13日條），被守備中部地區的副將（協台）樂文祥帶走，「失去這條狗，似乎使得異鄉客少了排解寂寞的管道；真不知老美為何未將薯條要回？『烏坵狗過台灣』的記載不再出現日記。而不論薯條是公是母，牠的後代顯然應該已遍佈彰化地區；除非喜愛美食的老樂另有盤算，他該不會把牠當成『熱狗』吧？」（1878年1月9日條註釋）。第一版寫自序之時，曾和內人帶著狗兒子Toro、狗女兒Yuki驅車前往踏查，而今Yuki仍相左右，Toro卻早已走丟。每每想起，痛徹心扉。

修訂版在即，在此感謝：中油公司同意筆者使用翻拍自該公司的老照片；老同學陳綠蔚、廖榮隆、林明福各方面的支持；Sophia幫忙緊急添購中、英文版《馬偕日記》，得能在三校即將交稿之際，及時讓全書有更正確的解讀；五南文化事業楊榮川董事長、楊士清總經理的再度青睞；台灣書房出版公司編輯群蘇美嬌、蔡明慧的辛勞；亦師亦友的王壽來、東年、翁佳音之長期關照提攜；小玉、Yuki的分享或苦或樂。

永懷那段陋巷歲月。是爲序。

▲美地質學家施幹克博士（Dr. Schenck）（左排後四）率美國專家參觀礦坑A2井（1952）。（中油，陳政三翻拍）

▲出礦坑A2井井架組立

▲出礦坑參觀油井附近（陳政三攝）

▲開礦村內大都是中油廠房，原來的出礦坑村應在本地西南邊山中（陳政三攝）

▲開礦村中油二氧化碳脫除廠

▲光復初期出礦坑中油辦公室（油礦開採陳列館展示照片，陳政三翻拍）

▲老君廟第一座煉油槽（中油，陳政三翻拍）

▲老君廟油田的老油人（中油，陳政三翻拍）

▲老君廟之採油裝置（中油，陳政三翻拍）

▲甘肅老君廟第一號深井（中油，陳政三翻拍）

▲出礦坑日治時代第一號古油井碑。（陳政三攝）

▲1955年山仔腳一號井試油氣實況——再次的「中美」合作（中油提供，陳政三翻拍）

▲郇和很早即提到台灣產石油，圖為中油出礦坑參觀油井（陳政三攝）

▲（錦水頁岩）地下岩層石油氣儲窖施工（中油，陳政三翻拍）

▲台灣首座海域油田——新竹外海長康油氣田（中油，陳政三翻拍）

▲（上）鑽井工程（下）陸上鈷井（中油，陳政三翻拍）

▲（頓鑽機動力所用之）蒸氣鍋爐（中油，陳政三翻拍）

▲出礦坑鑽油設備（陳政三攝）

▲中油出礦坑參觀井

▲出礦坑一〇七號井夜景（陳政三翻拍）

▲（左）出礦坑78、76、77、83號井（由左至右），（右）大肚山一號井（取自《中油史料影輯》，陳政三翻拍）

▲開礦村後龍溪溪北山上，應是清光緒年間第一口油井所在。（陳政三攝）

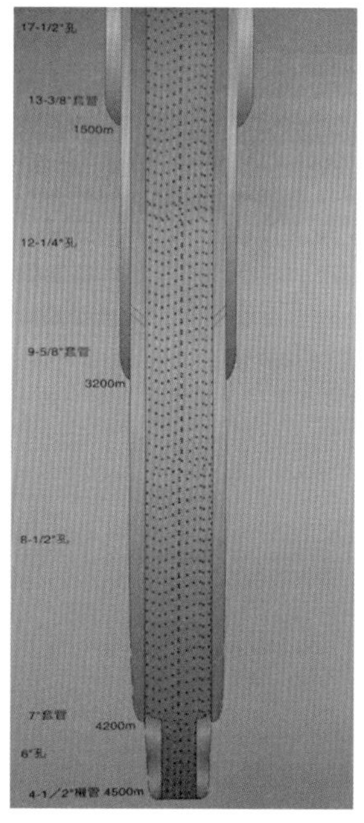

▲出礦坑141號井內剖面圖（中油《上山下海鑽井忙》，陳政三翻拍）

▲（中油）出磺坑雕像（陳政三攝）

▲一〇六號油井說明

▲公司寮港（今龍港）（位於後龍溪出海口南岸）（陳政三攝）

▲石油的形成（中油，陳政三翻拍）

▲油礦開採陳列館鑽井設備模型（陳政三攝）

▲各式鑽井套管及配備器具（陳政三攝於「油礦開採陳列館」）

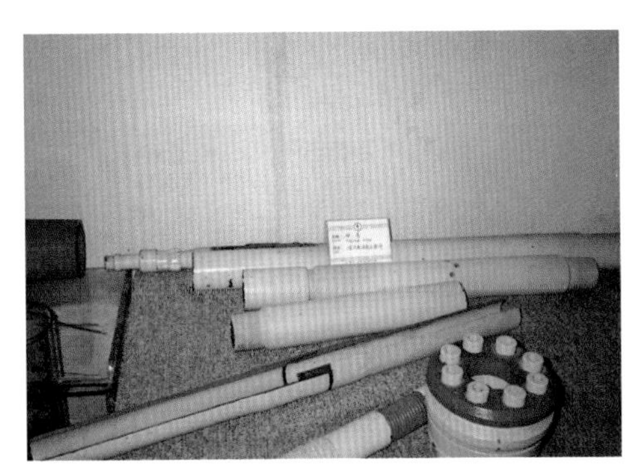

▲油管（陳政三攝）

▲（陳列館內）各式鑽頭（陳政三攝）

▲中油鑽井船模型（陳政三攝）

▲出礦坑鑽油設備

▲油田地區地質構造（中油，陳政三翻拍）

▲天工開物圖：古代鑽鑿石油、礦物情形（陳政三翻拍）

▲ 泥漿抽機（陳政三攝）

▲ 出礦坑抽油泵（陳政三攝）

▲（上）中油向法國租用第一套電測設備（1958）
（下）震波測勘（中油，陳政三翻拍）

▲2011文建會文資總處「青年論壇」出礦坑組學員：嚴致翎、簡佩茹、呂友文、林慧青、葉峻鳴、葉亦嚴（由右至左；陳政三攝）

▲「青年論壇」學員製作別出心裁的「出礦坑大富翁」遊戲（陳政三翻拍）

▲「青年論壇」出礦坑組學員製作的礦場解說圖（陳政三翻拍）

楔子

光緒元年（1875）十一月，接掌福建巡撫的丁日昌，執意在台灣興建鐵路、架構電報線、以西法開挖煤礦、鑽取石油，台灣在他的現代化觀念帶動下，有了一番新貌。美國技師簡時（A. Port Karns, 1840～1920年代），以及助手絡克（Robert D. Locke, 1850～1943）就是在這種大環境下，於1877（光緒三）年11月底來台。可惜丁日昌因推動新政受到各方掣肘，已在同年8月託病、請假返鄉，1878年春被迫短暫回任，但已無心政務，而於1878年5月正式卸任。政局動盪之際，美國技師發現他們花在旅行、等待的時間比實際工作時數長太多了，從美國故鄉出發，到台灣實際開始工作，足足拖了四個月，讓他們見識到五千年官場文化所孕育出的工作效率。加上又有令他們頭皮發麻的獵頭族、吃得膽戰心驚的飲食衛生、不良的工作與生活環境、人力不足的窘境、後勤欠佳的挫折等，終於倦勤並引爆了思鄉病，遂於合約滿一年後，不再續約，而於1878年11月初離台。他們的離去，宣告位於苗栗縣公館鄉出磺坑、這處清國第一座機器鑽挖油井的廢棄。絡克雖是美國油鄉出身的大老粗，所受教育不高，但卻爲我們留下長達一年三個多月的《出磺坑鑽油日記》，十分寶貴；相對於絡克，除了收於《清季臺灣洋務史料》（台銀文叢278種）4～28頁四件、《李文忠公選集》（文叢131種）199頁一件，共5件奏摺，以及沈葆楨與李鴻章的通信以外，目前卻仍找不出當時留下的其它中文相關記載，這更凸顯此日記的重要性。

Prologue

Ding Richang inaugurated as Governor of Fujian Province in November 1875. He persisted in constructing the railroad, erecting the telegraph line, excavating the coalmine and drilling the petroleum in Formosa (currently Taiwan). With this modernized idea, Formosa had had a new look. Two American oil technicians, A. Port Karns and Robert D. Locke, were hired by Fujian government and arrived at Taiwanfu (currently Tainan) in the end of November 1877 (the third year of Guangxu Emperor).

Yet, Governor Ding had not been smoothly in promoting his New Deal, which had been received all quarters of impediments and oppositions. He malingered, asking for sick leave, and returned to hometown in August 1877. Though Ding was forced to return to his post shortly and temporarily, yet he was unwilling to carry out his task as a Governor. He resigned officially in May 1878.

During the turbulent stage of political situation, Messrs. Karns and Locke found that they had spent too much time in traveling and waiting, much more than their working hours. It took 4 months from the day they left hometown until they started to work in Formosa. They understood Chinese 5,000 year tradition and inefficiency bred by government culture finally. Moreover, there were fearful headhunters, not hygienic diet, lousy work and living conditions, short of manpower and set back of poor logistics, etc. So they wearied from work and wanted to go back to the United States. They decided not to extend contracts with Chinese government after expiration and left Formosa at the beginning of November 1878. This announced the temporary end and abandonment of the first machine-drilled oil well at Chhut-hong-khinn (in Gongguan Town, Miaoli County).

Although Robert Locke was an uneducated person from American oil town in Pennsylvania, yet his habit of keeping a diary reserved a 15 month-long precious record for Taiwan. On the contrary, there are only few documents concerned left including 1 document in *Li Wenzhinggong Anthology*, 4 documents in *Historical Data of Taiwan Foreign Affairs in Qing Dynasty*, and correspondences of Shen Baozhen and of Li Hongzhang. Besides these, there is no other Chinese record on Taiwan oil drilling found so far. The fact highlights the importance of Locke's Diary.

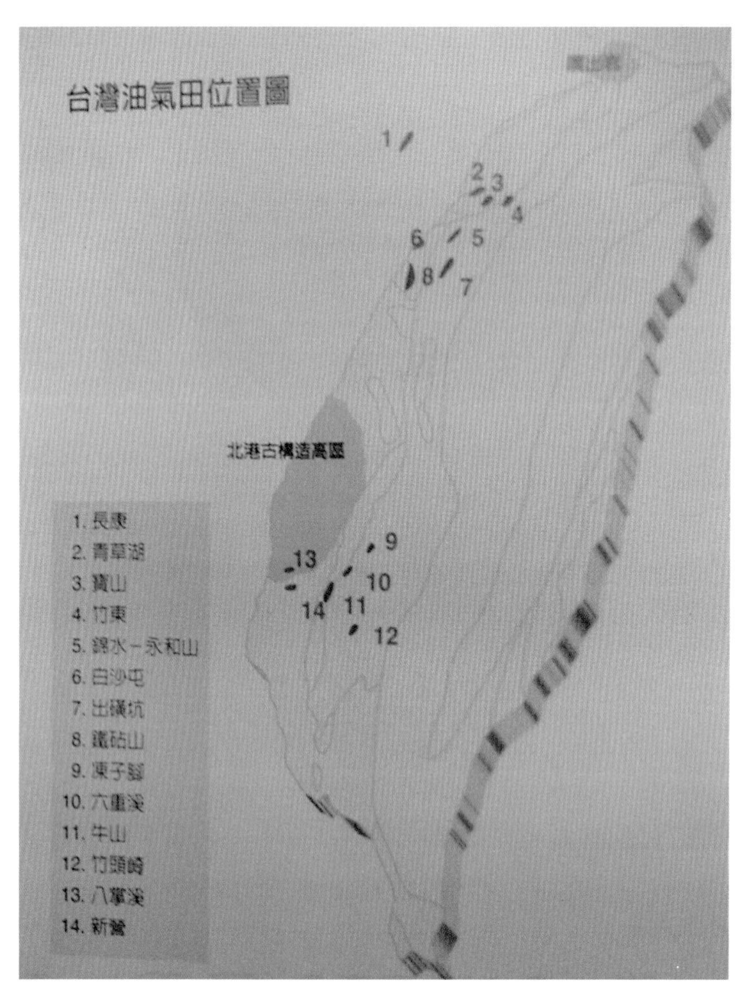

▲台灣油氣田位置圖（中油，陳政三翻拍）

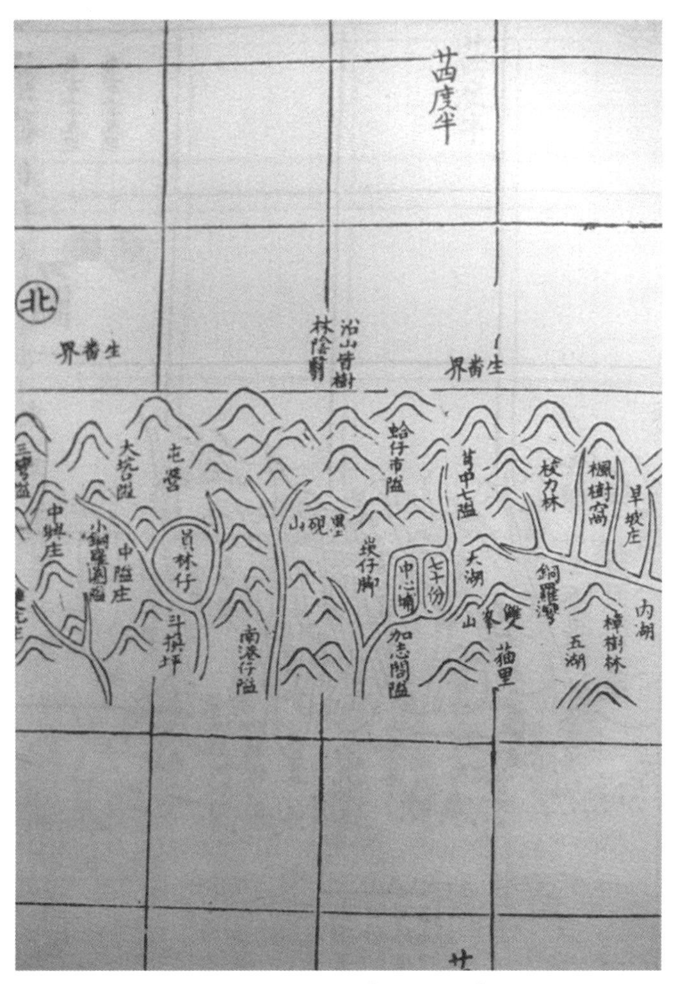

▲苗栗、哈仔市、銅鑼灣一帶（《淡水廳志》，陳政三翻拍）

首部曲

美國過台灣．心肝結歸丸

福爾摩沙初體驗

1859年，醉客上校（Colonel Drake）不知是用猜的、還是作夢夢見的、亦或占卜的，或是酒醉矇到的，反正他在美國賓州提塔斯維爾小鎮（Titusville）百般嘗試，終於鑽到了世界第一口油井，這座油井因之被命名爲「醉客油井」（the Drake Oil Well）。油鎮湧進不少採油客，因而興起，也培養出許多無師自通、土法煉鋼的油匠。

到70年代，東岸鑽不到油的油匠、西岸挖不出金的淘金客充斥，加上橫跨美國的火車鐵路工程也已完工，導致美國經濟不景氣，銀行相繼倒閉，失業率大增，遂興起席捲全國的排華風潮。多條限制、排斥華裔移民工作權的法案陸續通過，一船船載滿如鮭魚返鄉的華人船舶從西岸開出。出身油鄉的的簡時、絡克就是在這種環境下，因緣際會的搭上逃離美國的華工返鄉船，輾轉來到台灣。

1877年，主持台灣鑽油計劃的輪船招商局總辦唐景星，透過擔任駐美副公使的老同學容閎招募油匠。娶美國太太的容閎則再透過內兄（或內弟）凱洛（E. W. Kellogg）到油鎮找到了簡時，簡時堅持要有一位助手，於是又找到絡克。

9月4日，油鎮小車站擠滿送行的親朋好友，簡時、絡克帶著眾人的祝福「載譽出國」，搭上西行火車頭等車廂，直奔舊金山；而3萬美金購買的機器，爲了節省運費，則從紐約出港，繞了整個西半球輾轉來台。他們與心愛的鑽油機器此別，可要等到半年又十一天後才能再見。

經過8天的長途旅行，抵達舊金山，絡克匆匆拜訪舅父，即與簡時購買美國輪船「北京城號」（*City of Peking*）頭等艙船票，於9月12日出航，直奔日本，途中一路無話，只除了目睹一場病死華工的海葬。絡克回憶道，「華人由各個角落擁上甲板，擠在欄杆邊憤怒地高聲抗議。屍體終被拋入海中，揚起一片浪花，他們大聲齊嘆『哇！』（wagh），就結束了這段插曲，默默地回到各自的角落，彷彿沒有什麼事發生過似的」。

經過前後共22日的航行，於10月3日抵達日本橫濱港。下了北京城號，在日本各地前後盤旋8天，10月11日在長崎改搭一艘清國輪船，13日抵達上海。他們被安排住在上海第一流的亞士都飯店（the Astor House），前後住了33天，參觀了所有的名勝古蹟。每天由唐景星、布郎（Robert Morrison Brown）輪流陪伴，四處觀光、看戲、採購、上館子，到最後「看到中國菜就想吐」。還在百貨公司大肆採購，包括兩張床鋪、一套鍋、碗、瓢、盆、鏟、火爐樣式齊全的廚具、碗盆等，準備在苗栗山區展示這些時髦洋玩意。

在上海時，從某位《芝加哥論壇報》（*the Chicago Tribune*）記者口中首次聽到台灣居然有可怕的獵頭族，加上又看到一場遊行，行列中有人手持據稱是台灣原住民專用來砍下人頭的大矛斧，使得兩人頭皮發麻。簡時一度打退堂鼓，不去台灣了；由於不到台灣可是要賠償已花的巨額旅費，最後還是硬著頭皮、乖乖地照約行事。

11月14日終於離開上海，16日抵馬尾兵工廠，就住那裡，靜候代理巡撫葆亨撥冗接見。好不容易等巡撫賜完宴，又是5天過去了。20日清晨登上靖遠砲船，終於出發到久聞大名的美麗島；可天公不作美、風浪太大，到了24日早上才抵台灣府（台南）。兩位美國鑽油技師被安排住在孔廟旁台灣府學學生宿舍，有位會說一些英語（some English）的譯員陪著，見過「台灣土皇帝」道台夏獻綸、見過很多大小官員，也在11月28日與台灣通商局委員、即補分府鄭膺杰（Chang Ying-chieh）簽妥正式合約，清國（甲方）見證人是靖遠號砲船管帶（船長）葉阿富（Yip-a-Foo）；美方（乙方）見證人則爲英國駐府城領事館員何藍田（W. Holland）、當時兼管美國在府城的副領事事務。

中文合約規定：

臺灣通商局委員即補分府鄭 今與美國人簡時、絡克議立合同。事緣臺灣淡水所轄一帶地方所產煤油，前由容道臺在美國轉託商人布郎，招僱簡時、絡克二名來臺辦理開採油井事務，所有在美國原議合同係由布郎代訂。今到中國臺灣地方，將各事議明，應與通商局委員另立合同，彼此分執為據。

計開

一、此議單係上海布郎與美國簡時所立。

二、訂明簡時前赴臺灣辦理開採煤井事務，絡克係隨同辦事，均以一年為期滿。簡時每年薪水三千元，絡克每年薪水一千二百元，均按美國金錢價算。

三、簡時等前往中國之盤川歸官給發。飯食連用工人等項，每人每月從到臺〔起〕給洋銀五十元，歸其自理。所有住房、傢具等項，由中國官另給。

四、一年期滿之後，所有回國盤川均由官發給。若不滿期而簡時等自行告退，即不給矣；薪水亦應停止。

五、其薪水按三個月給一次。

六、簡時等到臺所有開井取油等工夫，務須自盡心力，認真辦理，不得無故曠工，一切須與中國地方官委員妥為商酌。所用學徒、工匠人等，並應盡心教導；如有不遵，告知委員分別責罰，不得凌辱。

七、訂明一年之期，由美國一千八百七十七年九月初四日，即中國光緒三年七月二十七日，簡時等動身之日起算。倘未到期而中國不用，仍照一年薪水發給。

八、簡時等如有患病，醫藥係自行料理。

九、簡時等在唐景星觀察處有領過薪水，應照扣算。

十、簡時等來中國盤川，已由唐景星觀察付給。將來一年工竣回國，每人應給盤川若干，應由唐觀察另行批定。

十一、如一年期滿，中國官員要再留在局辦理，薪水按月照給，不〔得〕以一年〔為例〕。

十二、簡時、絡克二人辦理煤油事務，〔必須〕工程〔完滿〕，交准中國官員為止。

▲美國鑽油技師簡時、絡克之中文合約（取材自Sampson Kuo書中資料）（陳政三翻拍）

英文合約原文（Original English Contract）：

An agreement made between Chang, Sub-Prefect and Deputy for Taiwan Board of Trade, Formosa, in the Empire of China, party of the first part, and A. P. Karns, an American citizen, U. S. A., a driller of oil well, and R. D. Locke, an American citizen, U. S. A., also a driller of oil well, under the superintendance of A. P. Karns, on the Island of Formosa, parties of the second part.

This agreement made in consideration of oil wells in the vicinity of Tamsui, in Formosa, through Yung Taotai of the Chinese Embassy to the United States, and R. M. Brown of Shanghai, China to engage A. P. Karns and Robert D. Locke as drillers of oil wells in Formosa. Agreement was being made at Titusville, Pa., U. S. A., between R. M. Brown and A. P. Karns and Robert D. Locke, parties of the second part, to replace by a similar agreement of the same import be made in China between said parties of the first and second parts as follow：

An agreement by and between R. M. Brown of Shanghai, China, and A. P. Karns and Robert D. Locke, both of Titusville, U. S. A. It is agreed that：

(1) A. P. Karns shall receive a salary of three thousand dollars（$3,000）gold per year, and Robert D. Locke one thousand two hundred（$1,200）as drillers of oil wells on the Island of Formosa.

(2) Also the travelling expenses of Karns and Locke to the Empire of China shall be paid. On arrival in Formosa an allowance of fifty dollars（$50）per month for board and servant shall be made. Lodging and furniture shall be provided.

(3) Also the return passage tickets to New York of Karns and Locke shall be paid, unless any of them breaks this agreement by leaving the service of the Petroleum Company without their consent, within one year from the date of leaving Titusville, the time this engagement began.

Salaries shall be paid up to such date.

(4) Also the salaries of Karns and Locke shall be paid as often as once in three months.

(5) Karns and Locke agree to do all in their power to faithfully promote the interests of the Petroleum Company and superintend oil wells in Formosa. On these matters, they will consult with the Chinese Authorities or its deputies. They agree to instruct or teach the workmen and laborers as far as their experience and abilities go without reserve. In case disobedience or, misconduct is found on the part of the laborers, they are to refer this to the deputies to punish the offenders and not undertake themselves to inflict the punishment

(6) It is mutually agreed that Karns and Locke will remain in the service for one year, salaries commencing September 4, 1877. If the Petroleum Company shall discharge one or both, he or they shall receive the full salaries as agreed upon for one year.

(7) It is agreed that in case of illness, Karns and Locke shall find themselves doctors and medicine.

(8) Also that the advance made by Tong Taotai to Karns and Locke for salaries shall be deducted here.

(9) Travelling expenses of Karns and Locke to China having been paid by Tong Taotai , and after the expiration of one year their return passage tickets and travelling expenses to New York shall be provided for by Tong Taotai.

(10) It is agreed that after the expiration of one year if the Chinese Authorities should consider it expedient for Karns and Locke to remain longer in the service in order to complete the whole work,

salaries shall be paid monthly at the same rate of $250 and $100 respectively per month but shall not be paid by the year.

(11) It is agreed that Karns and Locke shall remain in the service as drillers of oil wells in the Island of Formosa until the machine shops, buildings and drilling work of the Petroleum Company be brought to entire completion, giving satisfaction to the Chinese Authorities.

In witness whereof we have here unto our hands and seals at Taiwanfoo this twenty eighth day of November 1877, corresponding to Kuang-hsu 3rd year, 10th moon, 24th day.

(signed by)
Chang (in Chinese), Sub-Prefect and Deputy for Taiwan Board of Trade

Witnessed by
Yip-a-Foo (in English), Captain of H.I.C.M.S. "*Tsing Yune*"

(signed by)
A. P. Karns
R. D. Locke
Witnessed by
The undersigned
W. Holland, Assistant in Local charge of U.S. Acting Consulate

註：

1. 中文合約原件各項開頭均標上「一」字樣；編號係筆者加入。
2. 中文條文〔〕內之字，表示拓印本字跡模糊難辨，筆者依據字跡、語意推測而成，惟不敢確認是否爲原字樣。
3. 中文全約無標點符號，均由筆者加入。

11月29日，包含官吏、士兵、翻譯、苦力，還有老美請的廚子、僕人各1名，更有官兵隨行的眷屬，一行共約75人終於從府城出發，沿途在茅港尾（台南市下營區茅港里）、嘉義城、莿桐巷（雲林縣莿桐鄉莿桐村）、彰化城、牛罵頭（清水）、通霄過夜，12月7日抵達後壠（苗栗縣後龍鎮）。打從在彰化城遇到喜歡美食的副將樂文祥，往後旅途，駐防地方的小軍官爲了巴結陪伴老美同行的頂頭上司老樂，特地大擺筵席，他們因之沾光，開始吃香喝辣的，還吃到魚翅大餐、二十八道菜的盛宴。

12月8日到後壠港，發現這個將運機器上岸的港口水深不夠。9日離開後，沿後龍溪谷朝東南方向行30華里（約17.3公里），在離油泉5里處溪洲庄（公館鄉福德村內小聚落）過夜。10日，溯後龍溪上游，到5里外的出礦坑油泉探勘，發現，「出油地點夠寬敞，擺得下鑽油機器，並可搭蓋一間工寮」。於是老美心情愉快地返村，陪老煙槍官員老王抽鴉片。此時，已是離開故鄉三個月又六天。

正待好好工作、大展身手，突然接到調召他們返回府城的命令。不得已，12月12日從油村出發、打道回府，當晚夜宿葫蘆墩（豐原）；13日抵彰化城，招來剃頭師，「他連我頭頂的毛髮也想一起剃掉，但我堅持那個地方可是不讓人剃光的」，絡克寫道。由於這次不愉快的理髮經驗，使他從此不信任台灣的剃頭師，兩個禮拜後，他與簡時在不得已的情況下，還曾相互剪頭髮；之後，甚至可能長達七個半月未再剪過頭髮。

15日從彰化城出發，大致還是沿著來時的路徑南行，18日趕抵府城，發現被招回之因居然是夏道台爲了讓他們「在府城好好過個陽曆新年」！聞言令老美爲之氣結。但他們可是辜負了通曉人情世故的道台一番美意，兩個禮拜的特別假，除了不能免俗的拜會官員、賜宴吃飯，其它只是閒逛、購物、在家休息、互相剪頭髮，有次外出返回住處，發現居然遭小偷，飛掉煮熟的雞鴨各一隻。連平安夜、聖誕節，也未參加府城洋人圈的聚會；倒是拜訪了英國外交官何藍田，還向德馬太醫生（Dr. Matthew Dickson）多買些奎寧備用。

簡時在一封寄給美國同事的信中，陳述他對府城與台灣的印象：「此城人口約5萬，住在高25呎、厚12呎、周圍6哩長的城牆內，城牆夠寬、上可跑馬。氣候十分暖和，陰涼處介於華氏75度到80度之間（攝氏23.9～26.7度），人們整年都可睡在戶外。本島農產品很多，有鳳梨、橘子、柚子、香蕉、梨、甘薯、甘蔗、花生，以及煙草等。整體而言，是一個美麗的地方。」

體貼的夏道台送他們一座鐘（還好老美無太豐富的「送終」聯想）、一盞油燈、很多餐具；知道老美不敢吃不太衛生的豬肉，還遣人送來雞鴨各兩隻、四分之一隻山羊肉，供作新年加菜之用。1878年元旦，他們領到「紅包」——各50美元等值的本地通行銀元，那是兩人一個月的生活津貼。領到錢，也結束了府城之旅，次日、也就是在他們離鄉近四個月，束裝再行北上。

1877年（光緒三年）

1877年9月4日

〔與同伴簡時〕搭火車離開故鄉提塔斯維爾（Titusville, 位於賓夕凡尼亞州西北，產油小鎮）赴舊金山。夜宿伊銳鎮（Erie, 賓州）紐頓先生（Mr. Newton）的家。

September 4, 1877: Left Titusville for San Francisco stopped one night at Erie with Mr. Newton.

❶ 兩名美國技師之名，係根據中文合約上所用的名字。

❷ 全文字句後的（）為加註，字句前、後〔〕內的字句，係為使文意更清楚而加入，原文並沒有。

9月5日

凌晨3點離開伊銳鎮，晚上7點抵達芝加哥；當天晚上10點離開芝加哥，6日晚上7點抵達某地〔（愛阿華州）塔其歐溪（Tarkio Creek）〕，夜宿溪畔小鎮（按可能是Stanton）。

September 5: Left Erie 3 A M (A.M.) arrived in Chicago 7 P M (P.M.) Left Chicago 10 P M arrived at ? 7 P M on the 6th stoped one night.

❶ 英文用字係根據絡克原稿，他受的教育不多，所以有許多不合文法之處。如stoped正確用法應是stopped；A M與P M應為A.M.或a.m.及P.M.或p.m.；中間斷句部分，他常省掉句號，開頭字母也經常未大寫。本書依據

其原始寫法呈現。文中有？者，係由吉丹斯博士（Dr. Paul H. Giddens）整理時加入。

❷ 愛阿華州塔其歐溪（Tarkio Creek）及Stanton，為筆者根據前後行程之推測。

9月7日

上午9點抵達〔愛阿華州〕斷崖會議鎮（Council Bluffs, 原文寫爲Counsel Bluffs）。由於〔橫跨密蘇里河的〕橋樑坍塌，只好到下游20哩處搭船過河〔到對岸內布拉斯加州的俄馬哈市〕。下午3點，〔搭火車〕離開俄馬哈（Omaha）。

9月9日

傍晚5點，車抵〔猶他州〕歐格登（Ogden）；晚上7點，離開歐格登。

9月11日

〔經過8天橫越美國的長途旅行〕本日抵達舊金山；投宿普瑞絲蔻旅館（Prescott House）。

September 7: Arrived at Counsel Bluffs 9 A M bridge gone had to go down the river 20 miles and cross on a boat Left Omaha 3 P M arrived at Ogden on the 9th 5 P M Left Ogden 7 P M on the same day & arrived in San Francisco on the 11th Stopped at Prescott House.

❶ 7日之日記包括9、11兩日，可能是這3天來的「隨手記」。以下也經常有某天日記記載多天之事。

❷ 這趟火車之旅不便宜；如依據次年10月25日，台灣道台夏獻綸支付他們返鄉從舊金山到紐約的旅費，每人180美元來看，他們搭的應是頭等車廂。

9月12日

拜訪舅父哈內柏（Uncle Hannable）。〔與簡時〕購買赴上海的船票，〔每人〕美金372元。中午12點整，搭〔美國輪船〕「北京城號」（*the City of Peking*）出發。

September 12: Went to see uncle Hannable Bought ticket for Shanghai $372.00 Left 12 noon on the City of Peking.

❶ 絡克母親的娘家在舊金山；但隔年12月，由台灣返抵美國，再度探視哈內柏時，1878年12月12日日記將舅舅的名字寫為Hanables；13日日記寫成Hanable；不知何者才正確？絡克從小就在油鎮提塔斯維爾打混，12歲當鑽油廠的鍋爐小弟，13歲即自行創業、煉油，每天可生產4加侖、約2美金的石油，可知他大概只受過初等教育；從他的日記寫法、拼音錯誤百出，也可看出他是屬於大老粗之類的人物；但卻為我們留下來台近一年、以先進技術鑽取清國第一口油井、也是台灣第一口深入地底120公尺油井的寶貴資料。

❷ 簡時與絡克買的船票應是頭等艙，票價每張高達372美元；但次年11月中旬返美搭同艘船，清國當局支付他們由香港到舊金山船票每人只270美元，可能是商務二等艙的票價，不過為了省錢，買的卻是經濟艙（參閱1878年10月25日、11月15日日記）。前後的差異，應是去程被奉為上賓；但他們一年契約期滿後，不肯再續約，返程自然遭到「降格」的待遇。為了省下盤纏，買的是經濟艙。

❸ 英文船名目前用法採斜體字。

9月15日

傍晚6點，輪船引擎故障；次日凌晨2點才修復。

September 15: Engin broke down 6 in the evening did not get started until 2 in the morning.

9月24日

航過〔東經〕180度子午換日線，日期自動跳至25日。

September 24: Crossed the 180 degree line jumped the 25th.

10月3日

〔經過前後共22日的航行〕抵達日本橫濱港（Yokohama）。3名華人在航行途中死亡。

October 3: Arrived at Yokohama three Chinese died on the way from San Francisco.

❶ 根據絡克晚年接受故鄉《石油與瓦斯日報》（*The Oil and Gas Journal*）記者馬瑞（F. F. Murray）訪問，他表示船上擠滿因美國排華政策而返鄉的2,000名左右華工，當一名死在途中的華人遺體將海葬時，「華人由各個角落湧上甲板，擠在欄杆邊憤怒地高聲抗議。屍體終被拋入海中，揚起一片浪花，他們大聲齊嘆『哇！』（wagh），就結束了這段插曲，默默地回到各自的角落，彷彿沒有什麼事發生過似的」。這篇訪問刊於該報1931年10月1日，篇名〈一八七七年在清國鑽油的提塔斯維爾人〉（Titusville Man Drilled in China in 1877）。

❷ 絡克提到的華工，曾是淘金客、鐵路工，或洗衣工，但隨著1869年貫穿東西海岸的鐵路竣工；70年代初期加州金礦瀕臨枯竭；舊金山股市崩盤；銀行倒閉等，造成美國經濟不景氣，失業率大增。絡克、簡時肯千里迢迢到台灣打工，可能與當時工作難找有關。美國各政黨高喊「黃禍！」「華人滾回去！」等口號，其中最偏激的是梁啟超在〈記華工禁約〉乙文提到的「加州工人黨」領袖丹尼斯·奇亞尼（Dennis Kearney），他在舊金山港口西側沙地（sandlots）成立該黨，故又被稱作「沙地黨」，所推動的著名排華運動也因此被稱為「沙地運動」（the sandlots campaign）。

10月4日

改搭「東京瑪利亞號」（*Tokio Maria*）、原來的美國輪船「紐約號」（Old *New York*）離開橫濱。6日抵神戶（Kobe, 原文記爲Kabu）。

October 4: Left Yokohama on the Tokio Maria (Old New York) arrived at Kabu on the 6th.

❶ 「紐約號」原屬「美國太平洋郵輪公司」（the Pacific Mail Steamship Company），1874年（同治十三年）日本發動「台灣出兵事件」（即「牡丹社事件」），原已承租這艘噸位龐大、載量可觀的輪船當運兵、載補給品之用；但當時堅決反對日本攻台的英國駐日公使巴夏禮（Harry S. Parkes）向美國駐日公使平安（John A. Bingham）施壓，促使平安下令禁止該船租與日本。絡克的日記透露該船至遲在1877年已改掛日本旗幟了。

❷ 另外，他使用Tokio而非Tokyo稱東京，是當時的慣用法；隨日軍來台、採訪牡丹社事件的《紐約前鋒報》記者豪士（Edward H. House）在《征臺紀事》（*The Japanese Expedition to Formosa*）（1875）書中，也使用Tokio稱東京。

10月6日

由神戶赴京都（Kyoto, 原文用Kiota）。

October 6: Went to Kiota.

10月7日

返回神戶。

October 7: Went back to Kabu.

10月8日

凌晨4點，船離神戶，航經〔本州、四國間〕風景秀麗的〔瀨戶〕內海。

October 8: Left Kabu 4 A M Went through the inland sea beautiful scenary (scenery).

註 「風景」應拼為scenery.

10月10日

船抵〔九州〕長崎（Nagasaki, 原文爲Nagasaka）。上岸參觀該城。

October 10: arrived at Nagasaka went on shore & looked through the town.

10月11日

〔在日本前後停留8天，改搭一艘清國船〕離開長崎〔航向上海〕。

註 據絡克晚年接受訪問表示，離開長崎時搭乘的是清國船。

10月13日

下午4點，抵達上海。在這個古都滯留一個月（按前後33天，至11月14日離開），到處蹓躂、聽戲，參觀所有的名勝古蹟。

October 11: Left Nagasaka and arrived at Shanghai on the 13th 4 P M stayed in Shanghai one month went to through the old city to Chinese thetre & all the places of interest.

❶ 根據吉登獅（Paul Giddens）1940年訪問絡克，他與簡時被安排住在上海第一流的亞士都飯店（the Astor House）；每天由唐景星（Dick Sine）、布郎（Robert Morrison Brown）輪流陪伴，四處觀光、看戲、採購、上館子，到最後「看到中國菜就想吐」。

❷ 一位同船到上海的《芝加哥論壇報》（*the Chicago Tribune*）記者告訴他

們，他也要與朋友到「獵頭族」（head-hunters）居住的台灣深山採訪。這是他們第一次聽到台灣居然有可怕的獵頭族。加上，在上海又看到一場遊行，行列中有人手持據稱是台灣原住民專用來砍下人頭的大矛斧，使得兩人頭皮發麻。簡時一度打退堂鼓，不去台灣了；但絡克卻執意前往履約。由於不到台灣可是要賠償已花的巨額旅費，簡時最後還是硬著頭皮、乖乖地照約行事。可惜絡克未書明該記者名字；據現有資料，似乎也沒有這號人物來台採訪。待查。

❸ 唐景星（廷樞）（Dick Sine），廣東人，幼入「馬禮遜學堂」（the Morrison School），與容閎同窗共讀，未赴美國當小留學生，1863年進怡和洋行（Jardine, Matheson & Co.）當譯員，後升至總買辦，因此致富。1873年出任改組為商辦的「輪船招商局」總辦。1877年，福建巡撫丁日昌（1823～1882）為開發台灣石油，商得同鄉好友唐景星協助。唐乃透過在美照顧小留學生、兼任駐美副公使的容閎（Yung Wing, 1828～1912）幫忙；1875年容閎娶美國太太Mary L. Kellogg（1851～1886），遂請內兄（或內弟，待考）凱洛（E. W. Kellogg）到油鎮聘僱簡時、絡克兩位鑽油技師，並購買了3萬美金、一式兩套的鑽油設備。

❹ 布郎（1846～？）之父老布郎牧師（Rev. Samuel R. Brown, 1810～1880）曾是唐景星、容閎在馬禮遜學堂的校長，該校由1807年抵廣州傳教的英國傳教士馬禮遜（Robert Morrison, 1782～1834）在澳門所創；為紀念創辦人，老布郎乃以Robert Morrison當作兒子的名。出生於香港的小布郎從羅杰斯學院（Rutgers College）畢業後，返華做生意，後入九江海關擔任一等鈐字手（tidewaiter），

▲唐景星（唐廷樞）

1875年唐景星把布郎調到上海，成為他個人的特別助理。布郎約於1877年6月左右，曾赴美與簡時、絡克簽署聘僱草約（已散佚）。有趣的是，絡克於1878年底返美，詢問外祖父，發現在遠東鼎鼎大名的馬禮遜牧師居然是外公的Cousin（堂或表兄弟），換句話說，絡克要喊他舅公。世界真小。

❺ 馬禮遜為英國人，倫敦佈道會（London Missionary Society）牧師，1815年為了傳教，在麻六甲創辦、主編近代第一份中文報刊《察世俗每月統計傳》（Chinese Monthly Magazine）；1833年，在廣州創辦近代最早的中文雜誌《東西洋考每月統計傳》（Eastern-Western Monthly Magazine）；1841年，創辦香港首份英文報刊《香港公報》（Hongkong Gazette）；並曾為英文《廣東紀錄報》（Canton Register）、《中國文庫報》（Chinese Repository）執筆，堪稱近代中國新聞史的開創者之一。早期洋人稱玉山（又稱八通關山）「Mount Morrison」，據云係為彰顯他的貢獻而命名。根據W. Campbell（可能是甘為霖）的說法，係1844年英國Admiral Collinson為紀念1807來華傳教的馬禮遜牧師（Rev. Robert Morrison）而命名。1897年（明治三十年）明治天皇頒旨，將台灣最高峰玉山更名為「新高山」。W. Campbell, Mount Morrison, Formosa, The Chinese Recorder, Vol. 26, No. 7, pp. 333-334, also in: 張秀蓉（Chang Hsiu-jung）編, *A Chronology of 19th Century Writings on Formosa*, pp. 377-379.

10月23日

將一箱禮物託交「貓頭鷹巢號」（*Calio Owl*）輪船運送回家。

October 23: sent a box home on the ship Calio Ow.

❶ 11月2日絡克在寫給妹妹的信中稱，「花美金150元，買下一套鍋、碗、瓢、鏟、火爐樣樣齊全的廚具；（未寫價錢的）兩張床鋪」。可以想見，他們應是在上海的大百貨公司採購的；有趣的是，這些洋玩意擺在

苗栗山區油井附近廟中、工寮內，不知呈現出什麼畫面？

❷ 他們在上海滯留前後33天之因，除了鑽油機器尚在途中；最主要是推動台灣各項新政建設的丁日昌，恰於8月中旬奉准「回籍養痾」，巡撫職暫由布政使葆亨代理，丁於1878年春短暫回任，旋於5月正式卸任，由吳贊誠接任福建巡撫。值此政壇動盪之際，加上閩浙總督何璟向以保守、反對新政著稱，所以唐景星可能在等候上級是否支持繼續鑽油的新指示，使得兩位老美傻等，也因此除了血拼（shopping）、看戲、觀光之外，無新鮮事可做，日記因此空白許多天。

11月14日

搭乘「大禹輪」（？，*Thayew*）離上海、赴福州。

11月16日

上午7點，抵達馬尾兵工廠（arsenal, 原文爲arsinal），與Mr. Healand同住。

November 14: Left Shanghai on the steamer Thayew for Foochow on the 16th 7 A M stopped at the arsinal (arsenal) with Mr. Headland.

註 馬尾兵工廠於1866年由左宗棠所創設，與馬尾造船廠、船政學校屬同系統。當時洋人稱馬尾造船廠為「Pagoda Anchorage」，係因附近山坡有座高聳寶塔。

11月17日

〔搭船沿閩江〕上溯到福州市，與黑吉先生（Mr. Hedge）共進午餐，他還介紹新罕布夏州蘭卡斯特來的羅杰斯醫生（Dr. Rogers）與我們認識。回兵工廠。晚間接到〔署〕福建巡撫葆亨的邀請函。

November 17: went up to Foochow City took tefin with Mr. Hedge was introduced to Dr. Rogers from Lancaster N H went back to arsenal in the evening received an invitation to visit His Excellency Bow the Governor at Foochow.

11月18日

星期日。赴福州見巡撫。先搭蒸氣拖船上行13哩，再改坐轎子走6哩，與巡撫（governor, 原文爲govinor）共進午宴。宴畢，照原路回到河邊，等候漲潮搭舢板船時，看到一具死嬰順流漂下。使喚替我們買到麵包的童僕再去買雪茄，晚間7點，他買回雪茄。8點半，返抵兵工廠。

November 18: Sunday went to Foochow to visit the govinor (governor) went up the river on steam tug 13 miles then went back to river while waiting for tide took ride in sampan see dead baby floating down the river sent boy after cigars brought us bread sent him back 7 got cigars got back to the arsenal half past eight P M.

11月19日

與Mr. Hailand（可能即是前述的Healand）在兵工廠區散步。下午，將行李先送上清國海軍砲船。

November 19: went through the arsinal (arsenal) took walk with Mr. Hailand in the afternoon sent baggage board Chinese gun boat.

11月20日

凌晨4點半，被苦力（cooley, 應爲coolie）叫醒，他說編號第九號的靖遠輪（*Tsing-Young*, No. 9）即將啓航。趕緊上船，但直到早上8點，船才離岸航往福爾摩沙（Formosa）台灣府（Tia-wun-foo）。下午3點，風浪太大，灣靠（閩江口、馬祖南竿西南）尖峰島（Sharp Peak Is., 又名川石島，原文記爲heigh head）。

November 20: Cooley (Coolie) came & woked (woke) us to go on board (half past four A M) say (said) that the Tsing-Young (No 9) was going to leave went on board but they did not start until 8 A M for Tia-wun-foo (Taiwanfu or Taiwnafoo) Formosa. 3 P M anchored at heigh head on account of bad wether (weather).

註 正確拼法：cooley→coolie; woked→woke; say→said; Tia-wun-foo→Taiwanfu or Taiwnafoo; wether→weather。以下如原文拼法有誤，直接在文中該字後括號內修正。

11月21日

風勢太強，天氣很壞〔，仍停留尖峰島〕。

November 21: Wether (Weather) bad very strong winds.

11月22日

凌晨2點，開往烏坵島燈塔。上午9點抵烏坵（Ockseu, 原文Ocksen, 可能是筆誤，或因字跡潦草而印刷錯誤），上岸參觀〔1872年建的〕燈塔。〔燈塔守護員〕葛林（Mr. Green）送我們一條叫「薯條」（Chip, 原文Chep）的狗。

November 22: Started for Ocksen light house 2 A M arrived at Ocksen 9 A M went on shore went through the light house Mr. Green gave us a dog name Chep.

11月23日

凌晨2點啓航，上午8點抵達Tiawun（可能是大安或內垵）。上岸參觀，是個很骯髒的小鎮。午夜12點離開Tiawun, 24日上午7點抵達台灣府。

November 23: Started for Tiawun 2 A M arrived at 8 A M went on shore small town it very dirty Left Tiawun 12 mid night (midnight) for Tiawunfoo (Taiwanfoo) arrived 7 A M on 24th.

註 絡克使用Formosa稱台灣，而用Tiawunfoo或Taiwanfoo稱台灣府（台南），Tiawun是一個距離府城輪船航程7小時的小鎮，顯見不是Taiwan（台灣）之誤筆。以此推測，該地可能有二：（一）位於中部沿岸，再依發音判斷，可能就是大安港（台中市大安區大安溪出海口）。明代稱大安港為海翁窟港（鯨魚港），清初稱螺施港，港深可容大船出入。道光以後，淤塞僅容小船。如是，只不知為何要在該地停留16小時？引擎故障？或避風關係？（二）如果Tiawun指的不是台灣島某地，那麼可能是航程時數約略相當、發音近似的澎湖西嶼島上的內垵了。澎湖向為海軍基地，這或許較能解釋為何停留那麼久之因。內垵西側外垵西式燈塔於1875年竣工，駐有海關洋雇員，惜絡克未提到這座燈塔，否則可增加解讀是否為該地的資料。（三）Sampson Hsiang-chang Kuo在他的博士論文（1981）*Drilling Oil in Taiwan: A Case Study of Two American Technicians' Contribution to Modernization in Late Nineteenth-Century China*, 頁88，認為可能是鹿港或北港。但發音相去甚遠。

11月24日

前去拜會道台〔夏獻綸〕，他外出不在官邸。與一位小官及靖遠輪管帶〔船長葉阿富〕午餐。下午4點，道台返家，但太忙不克接見我們，只差人帶我們到學生宿舍（可能是南門路孔廟旁台灣府學）過夜。

November 24: went to see the Taoti (Taotai) he was not at home take dinner with small mandarin & the Captain of No 9 Taoti came home 4 P M Taoti to (too) busy to see us sent us to stope (stop) over night with some

Chinese students.

11月25日

週日。上午10點往謁道台，等到中午12點才獲接見，他與我們共進午餐、邊吃邊聊了三個鐘頭。回到昨晚下榻處，行李已從砲船搬來，包括兩張床、整套廚具火爐、兩件大皮箱、兩件手提小背包，還有那隻狗狗……等共17件。

November 25: Sunday 10 A M went to see the Taoti waited until 12 noon then was shown into his presense (presence) took dinner talked business for three hours went to the house where we stayed last night found all our baggage has been brought up from the gun boat (gunboat) Our baggage consists of 2 beds complete one cook stove two trunks two hand satchels & a dog 17 packages in all.

11月26日

上街散步，天氣太熱，未走遠。一位將與我們同往油泉（oil springs）的〔王姓〕官員來訪。下午，兩位大官來訪，他們〔透過翻譯〕問了許多有關如何鑽取油井的事；我們想從他們嘴中得知油泉的一些事，但他們似乎不願多說。

November 26: Took a walk through the town but it was so warm that did not go far Received a visit from mandarin that is to go with us to the oil springs in the afternoon received a visit from two large mandarin (mandarins) who asked a great many questions about operating for oil; tried (tried) to find out some thing about the place where they find (found) this oil but could not.

註 兩位老美抵台後，官方先後派遣幾個譯員陪同。據1877年12月23日簡時寫給公司（the Gibbs & Sterrett Manufacturing Co.）的信（次年3月2日刊於故鄉的《提塔斯維爾前鋒晨報》（*The Titusville Morning Herald*），標題為「The Far East」，以下略為〈遠東函〉），「有四個會說一些英語（some English）的漢人陪著，所以我們大致還過得去」，這四位會說洋涇濱英語的人，可能包括譯員及僕人。本日的two large mandarin應使用複數mandarins。

11月27日

寫信給康寧（D. Cummings）。靖遠輪船長葉富（Yip Foo）來訪，說他正在準備我們與清國官方即將簽署的正式合約。聘僱隨行廚子、僕人各1名；但稍後他們折回，稱渠等無法長途跋涉，必須坐轎才肯隨我們〔赴苗栗後壠〕上任。告訴他們明天前來聽取我們的決定。

November 27: Wrote letter to D Cummings received from Yip Foo (Captain of the Chinese gunboat) who is making out some new papers between the Chinese & us Engaged cook & boy to go with us but they came back & said that they could not walk & that if they went we would have to hire sedan chairs for them told them to come back tomorrow & we would let them know.

❶ 靖遠輪船長在中文合約上簽葉富，英文合約見證人欄上簽的是葉阿富Yip-a-Foo。

❷ 簡時、絡克在府城請的編號第一號僕人（No.1 boy）阿三（Ah Sun），略通英文，不知是原本會說，亦或與他們朝夕相處、耳濡目染而學會？阿三一直陪伴他們。Ah Sun依據閩南語發音，可為「阿順」；不過筆者從小到大，一直被呼為「阿三」，因此姑且化身融入書中情境。

11月28日

寫信給西番（J. L. Seyfang）。〔簽署正式聘僱合約，之後〕中午，道台賞宴。道台及同知〔鄭膺杰〕給我們幾罐茶葉。決定僱用廚師、僕人，每月薪水分別爲13、10〔銀〕元，先各預付每人8元。送洗衣物已好，27件才5毛。

November 28: Wrote letter to J L Seyfang took dinner at the Taotia Taotia sub mandarin gave us box of tea hired cook for $13 per month boy for $ 10 to go with us to the oil springs washing came home paid 50 cts for 27 peaces paid the cook & boy $8 in advance.

❶ 根據中英文合約上的簽署日為本日，絡克居然漏記！清方簽署人為臺灣通商局委員、即補分府（sub-prefect）鄭膺杰，見證人是靖遠號砲船船長葉阿富；美方當然是由簡時與助手絡克簽署，見證人為英國駐府城領事館員何藍田（W. Holland）、當時兼管美國副領事事務；稍後何藍田於1890～92年間，出任英駐淡水署領事（Acting-Consul）。1877年4月中旬至1879年初，英駐台灣署領事（轄台灣府與打狗）為費里德（Alexander Frater）。

❷ 鄭膺杰，廣東清遠人，監生出身，光緒七年以前曾主持基隆煤務；十年二月出任澎湖廳通判；次年陽曆3月底法軍攻打澎湖，軍前逃亡，加上先前經手煤務弊端，被流放黑龍江。參閱《劉壯肅公奏議》，頁429～434；《澎湖廳志》，頁194。

❸ 台灣道台夏獻綸顯然在場，為何不是由他代表簽字呢？這牽涉到官場上的「對等」原則，堂堂「台灣土皇帝」怎可與「洋匠」平起平坐？類似簡時、絡克等禮聘的西洋技師，在丁日昌眼中也不過是，「我但僱洋人為工匠，工竣則洋人可撤；將來一面舉行，即一面學習，不過二、三年，當可自為製造」。理論上，沒啥不對；但清末的自強運動，有哪一樣不是虎頭蛇尾、草草結束的，遑論「師夷長技以制夷」。

❹ 根據合約，技師簡時年薪美金3,000元，助手絡克年薪1,200美元；每人每個月伙食、僱傭、雜支零用金可再各領50美元（洋銀）。付給廚師、僕人的薪水應不是美金，而是清國銀元，當時兌換率1銀元約值0.77美元，換算起來，廚師月薪10美元、僮僕7.7美元算是高待遇。對照當時班兵月薪才6～7兩，如領的是成色十足的海關兩，據1892年清國海關總報告，1海關兩等於1.07美元，士兵薪水也不過合美金6.42～7.49元之間；但如不幸領的是成色差的地方銀兩，如據〈1882～91年台南海關十年報告〉，每1海關兩可兌1.1137地方銀兩，那麼班兵的待遇就更差了；難怪他們士氣低落，還要兼差當地方角頭的保鑣，或做起生意、客串「檳榔阿兵哥」。據絡克晚年的回憶，苦力每天工資約美金1.8角，就更差了。往昔與現在一樣，捧洋飯碗的待遇總是較高。

11月29日

到英國領事〔館員何藍田〕處，確定我們簽署的合約沒問題後，返住處打點好行囊，準備出發。但卻出了差錯，僱給2位僕人的轎子遲遲未到，足足等了2小時到11點半，轎子還是不來，〔一行75人〕只好出發。經過40華里（li）路程，傍晚6點抵達茅港尾（Ung-kang-bay, 台南市下營區茅港里）。上午7點吃一碗飯當早餐；晚上7點，吃了一碗飯，配上四顆蛋做成的炒蛋（4 eggs fritters）。廚子買了兩隻雞，明早準備當早餐。

November 29: Ung-kang-bay After visiting the English consul to see that our papers were all right we ready to start for this place but through some mistake the chairs for our boys did not come waited 2 hours & had to start without them at half past eleven arrived here 6 P M after travelling 40 li 7 P M had bowl rice 4 eggs fritters the cook bought 2 chickens for one breakfast.

❶ 1英里＝1哩＝1.60935公里＝2.7938316華里（里）。40華里＝24.85公里＝

14.3英里。1華里＝0.62125公里＝0.3575英里。

❷ 根據英國駐淡水領事館員阿赫伯（Herbert J. Allen）1875年底，從北部到台南的旅行記，茅港尾有家舒適的官營客棧；絡克一行應即下榻於此客棧。請參閱陳政三，〈英國外交官亞倫的畢業旅行〉，收於《歷史月刊》205期。筆者原譯為「亞倫」，經查他的正式漢名為「阿赫伯」。

❸ 他們一行浩浩蕩蕩，究竟多少人？據1940年90歲的絡克受訪時表示，包括200名士兵（許多攜帶眷屬）、11名貼身保鑣（部分帶著老婆隨行）、150名苦力、2位工頭、5名官員、3位小吏、2位廚子、1名出納；再加上一位從彰化率50兵到油泉處的軍官，如不包括兩位老美、僕人阿三、士兵與保鑣的眷屬，光是上述已達425人，陣仗比道台出巡還大。人在面對採訪時總免不了加油添醋，他要不是吹大牛皮，就是年紀太大、記憶力衰退而搞混。簡時在〈遠東函〉的記載應該較為正確，「抵達彰化城時，我們一行共約75人」。

❹ 供僕人乘坐的轎子為何未到呢？很可能是小官吏搞鬼，根本未代僱。在那個時代，哪有下人坐轎的道理？小吏當然看不順眼。

❺ Ung-kang-bay係茅港尾的台語發音，早期外國人書寫台灣地名用的是當時最通行的閩南語拼音。

11月30日

早上7點進早餐，有米飯、豬肉，雞肉留到中午吃。僱到兩轎供兩僕乘坐，上午8點離開茅港尾，每約一小時停轎休息，下午1點（原文誤植爲下午7點）吃午餐。沿途田園平坦、耕種良好，主要作物有甘薯、稻米、甘蔗。經60里路，傍晚6點抵達，住進一處舒適的官邸。今晚，〔首度試著〕用筷子吃飯。

November 30: Kagee Left Ung kang bay 8 A M had rice 7 pork for breakfast took chickens along for dinner Procured two chairs for boys stoped (stopped) to rest about once an hour stoped for dinner 7 (1) P M & arrived at this place 6 P M after traveling 60 li The country so far has been level & good farming land the principal crops are sweet potatoes rice & shugar (sugar) cane tonight we are stoping (stopping) at a mandarin house I have good accomadation ate a Chinese supper with chop sticks (chopsticks).

❶ 當時嘉義知縣為吳鳳笙，1877年11月間甫由壽寧知縣調抵台灣；隔年即卸任，日期不詳。

❷ 午餐時間應是下午1點誤植為7點。

12月1日

整天在衙門休息。〔吩咐廚子〕煮些東西準備明天在路上吃；另買三隻鴨及三隻雛雞（chicklings, 原文誤植爲chickings）、肉羊（mutton）腿一隻，但我總覺得那是山羊（goat）腿肉。晚上，公堂審訊一位被控謀殺罪的嫌犯，審問一個半小時後，發監囚禁，等遠赴淡水公幹的知縣回來後再行發落。據說等知縣回衙，犯人馬上會被處死。

December 1: Stoped at the mandarin Yamen all day to rest & cook some things to take with us bought three ducks & chickings (chicklings) & a leg of mutton but I think that it is goat Tonight there was a man brought in charged with murder the hearing last about one half hour & he was put in jail to wait the return of the chief mandarin who is away at present at Tamsui They say that he will die as soon as the mandarin return.

12月2日

〔星期日〕換了另一批扛轎苦力，上午7點半正準備上路，但同行的一位官員〔老王〕尚未過足煙癮，直等到9點才出發。我的轎子起步太慢，跟不上隊伍，不得不把愛犬〔薯條〕丟出轎外，以減輕重量；但完全無效，轎夫整天都跟不上隊伍。走了50里路，夜宿莿桐巷（Cheng-long-hang, 雲林縣莿桐鄉莿桐村）小村〔客棧〕。

December 2: Were ready to start at half past seven but owing to one of our mandarins taking too much opium we did not get started until 9 A M My cooleys got behind at the start & I had to throw the dog out the lighten but it did no good they stayed behind all day Traveled 50 li & stopped at a small village This morning we started with a new set of cooleys (coolies).

12月3日

早上下著毛毛雨，8點從莿桐巷出發。橫渡笨港溪（river Peneham, 北港溪），以及叫做華武壠溪（great river Touvar Yangh, 濁水溪支流新虎尾溪）的大河。道路泥濘使得苦力舉步維艱、速度極慢，距離今夜落腳處5里路時，天色已暗，要僕人（阿三）到村莊商借火把，但風勢太強，很難點燃。終於在晚上9點抵達彰化城（Chang-hwa），在縣衙與地方官員共進晚餐。今天走了50里（28.8公里）。

December 3: Cheng-long-hang Started at 8 A M light rain crossed the river Peneham & the great river Touvar Yangh The walking was so bad that it was very hard on the cooleys and they went very slow when within five li of our stoping place night overtook us & we had to send the boy after torches but the wind blew them so hard that we had some trouble to make them burn. Reached Chang-hwa 9 P M & took supper with a mandarin at his yamen Traveled 50 li.

❶ Sampson Kuo（論文頁98）對照李仙得地圖上標的Favor Langh, 認為即是濁水溪，該溪曾被稱做番挖港溪（Fan-wa-kang-chi）；絡克可能聽錯、誤拼。

❷ 不論李仙得的寫法或荷治時期的Favorolang，指的都是虎尾；打開周璽修的《彰化縣志》彰化山川全圖，濁水溪出海口南端即有虎尾溪同流出海。而這條讓絡克將它與濁水溪混在一起的當時虎尾溪，今名新虎尾溪。

❸ 道光年間，彰化縣芳苑鄉三林港淤積，南方番挖港興起，地近濁水溪口。

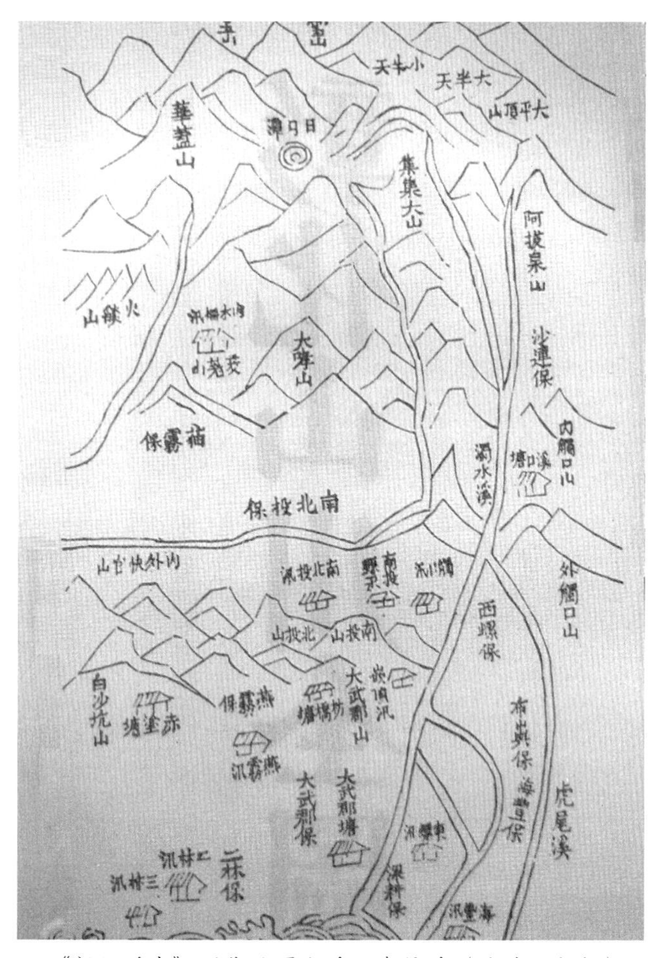

▲《彰化縣志》刊載的濁水溪、虎尾溪（陳政三翻拍）

12月4日

整天待在彰化縣衙，好讓同行官員休息、抽鴉片。與投宿的官邸屋主（知縣鍾鴻逵？）共進午餐，他送我們一包約重2磅的茶葉；由於他的朋友染患熱病，我們回贈一些奎寧（quinine）。晚上，那官員擺了我迄今吃過最豐盛的中式宴席，足足有二十多道菜。白天，吩咐廚子上菜市場買兩隻鴨、兩隻雞，供明天路上吃；途中白米取得方便，就沒買了。

December 4: Chang-hwa Stoped (Stopped) all day at the yamah (yamen) to let our mandarin rest & smoke opium took dinner with the mandarin of the house who gave us a package of tea about two lbs had our cook buy two cucks & chickens to take with us can get plenty of rice on the way gave the mandarin some quenine (quinine) for a friend of his who has a feaver took supper with the mandarin the largest Chinese supper that I have seen more than 20 different dishes.

註 當時彰化知縣為鍾鴻逵，廣東海陽人，1877年6月之前～1879年1月24日在任，因故奉旨革職。

12月5日

清晨6點，被一陣鑼鼓、絲竹聲吵醒，原來今天是農曆11月1日，人們正祭拜神明，神案前擺著三杯茶、三顆蛋、以及三牲。換另一組苦力接棒，上午9點半出發，一位〔樂姓〕副將（colonel）率20名左右的士兵隨行。走30里，在牛罵頭（Goomatao, 台中市清水區）過夜，與當地一位官員共進晚餐。此地是處極糟的小鎮。

December 5: Woked (Woke) up this morning at 6 A M there was drums fifes playing found out that this is the first day of the Chinese eleventh month & they were having a religious meeting & offering food to there (their) gods they gave three cups of tea three eggs & three ? of some kind of meat Started 9:30 A M with a new lot of cooleys (coolies) some soldiers about 20 & a colonel went 30 li & stoped (stopped) for the night at Goomatao Took supper with a mandarin This is a very bad town.

註 1878年8月22日日記，提到這位「上校」的名字為Lock Lie Len；依據「江西老表」不太標準的官話（北京話）發音，Lie Len近似「容軒」，推測可能是《苗栗縣志》（頁201～202）上載，於光緒元年（1875）曾署台灣北路營遊擊（資淺上校或資深中校）的樂文祥（字容軒，江西樂安人），光緒二年曾擒斬雞籠山土匪吳阿來建功；如是，他已因功升為駐紮彰化城的北路協鎮營副將（協台），統馭北路協中營（駐彰化）、北路協左營（嘉義營）、北路協右營（竹塹營）。

12月6日

上午8點半離開牛罵頭，到達15里處〔大甲〕，當地軍官（mandarin）請我們吃魚翅大餐。下午一路濕冷，再行25里，（在通宵）停轎過夜。駐守地方的軍官擺宴款待。

December 6: Left Goomatao 8:30 A M went 75 (15) li & stopped for dinner a mandaring (mandarin) had a very good dinner the principal dish was shark fins This afternoon has been very wet & quit (quite) cold after going 25 li stoped (stopped) for the night & took supper with the mandarin that is dealing with us.

❶ 中午、晚間佇足地，絡克均未寫出地名，筆者依據里程、有駐紮官員處推估出可能的地名。牛罵頭至大甲里程不可能有75里（約46.6公里），似為15里（約9.32公里）之誤。

❷ 途中小地方遇到的官員（mandarin）似應為由守備（正五品，資淺少校或資深上尉）、千總（正六品，約等同現在的上尉）、把總（正七品，中尉）、外委（可為正八品的外委千總，或正九品的外委把總，等同少尉及排副）、額外外委（士官長）等中下級軍官駐防的「汛塘」，故譯作軍官。防地大者曰汛（也兼管治安，類似目前的警察分局），兵力十幾、數十名不等；小者曰塘（派出所），士兵不到10人。以大甲守備署為例，道光七年起，設守備、千總、把總衙署各一，外委公所三間。他們擺宴款待來賓，並非巴結外賓，而是衝著頂頭上司、從二品的副將老樂。

12月7日

上午8點半出發，走了30里抵達後壠（Oulan, 苗栗縣後龍鎮），夜宿該地〔軍官的家〕。看到從油泉攜來的石油。

December 7: Started 8:30 A M went 30 li & stoped (stopped) at Oulan for the night see some oil that was brought from the oil springs.

註 絡克未交代較笨重的床鋪、火爐以什麼方式運送。合理推測，似應用牛車載運；但據簡時〈遠東函〉，「福爾摩沙無道路（roads），只有步行小徑（foot paths）」。由此可見，似乎仍是由苦力挑運。

12月8日

從後西行約1英里，到將運送鑽油機械上岸的〔後龍港〕港口，再搭小船出港，直到風浪太大、無法繼續前進方停止，測得水深才6呎（約1.83公尺）。返後，以冷雞肉、米飯當午餐，好奇的村民圍觀、有的甚至登堂入室，軍官命人用一塊紅布垂掛門口，以示此處乃官邸，並擋住「觀眾」視線，此舉收到令人滿意的效果。

December 8: went about one mile from here to see the harbor where our machinery is to land went out in a boat as far as we could go on account of rough water I found the water 6 feet deep came back and ate our dinner of cold chicken & rice The natives crowded around our house (& some of them came in) that the mandarin ordered a red cloth put over the door which is a sign that this is a mandarin house to keep them away which had the desired effect.

註 公司寮港位於今後龍溪出海口南岸，現名龍港，是日治初期，因後龍港淤塞，而興建的替代港口。

12月9日

星期日。以米飯、蛋當早餐，上午8點離開後，〔沿後龍溪谷朝東南方向〕行30里，在離油泉5里處〔溪洲〕過夜。晚上，獲邀赴宴。宴畢返住處，乖乖，屋子內外擠滿來看熱鬧的漢人，趕走他們的唯一方法就是吹熄油燈；待燈火熄滅，村民爭先恐後離開，彷彿後有魔鬼在追似的。今晚，打開自備的油罐，終於有了自抵此島後，最光明的一個晚上。

December 9: Sunday Left Oulan 8 A M after eating a breakfast of rice & eggs traveled 30 li & stoped (stopped) within 5 li of the oil spring was invited out to supper when we came back the house was full of Chinemen (Chinamen or Chinees) all the (only) way we could get them to go out was to put out the light then they went as if the devil was after them to night we opened our can of oil & had the first good light that we have had since we have been in the island.

❶ 溪洲似位於今苗栗縣公館鄉福德村，原先正確的位置似在後龍溪上游北岸，離溪岸1.44公里，是隔年他們第二度北上，從1月9日至8月22日，住了七個多月的地方。附近幾處油泉冒出的石油呈黃色，被誤為是硫磺，因此得名。絡克在回憶時，云鑽出的石油呈琥珀色澤；丁日昌光緒三年三月二十五日奏稱：「磺油……其色黃綠，氣味與洋油相埒。井之左右有十餘窟，亦有油浮水面。」目前以中油廠房為主的開礦村出磺坑，就緊鄰後龍溪南畔，在原先出磺坑附近。

❷ 粵籍客家人吳琳芳於嘉慶二十二年（1817）入墾石圍牆；吳昌和於咸豐初年（1851）入墾出磺坑地區；咸豐七年，邱苟組織「金長和」墾號十二股開闢出磺坑，他並將湧出的油泉出租牟利，不料卻惹來殺身之禍。詳1878年5月30日日記。

❸ Chinaman或Chinee有貶抑的意思；Chinese則為一般中性稱呼。

12月10日

溯〔後龍〕溪上行5里，來到〔出磺坑〕油泉探勘，出油地點夠寬敞，擺得下鑽油機器，並可蓋一間工寮。〔心情愉快地〕返村，陪官員〔老王〕抽鴉片。午餐：米飯、鵝肉、甘薯。晚餐：甘薯、鵝肉、米飯。

December 10: went up the river 5 li to see the oil springs looks very well located a place to put down a well & to build a house came back & smoked opium with the mandarin had rice goose & sweet potatoes for dinner has potatoes goose & rice for supper.

註 第一處開鑿的出磺坑油井在溪南。初版解讀為油井在「溪北」，修訂版更正為溪南之因，是由於完整的《馬偕日記》之刊行，記載了油村名「溪洲」，筆者乃重新思考、反覆推敲後，所做的修正。另，假使在溪北，不可能未留下任何遺跡。

12月11日

從小村到溪流下游1哩處，探勘開路地點，以便運送機器上山。午前11點返村，一直待在屋內。首次吩咐廚子試做西點麵包。

December 11: Went down the river about one mile to see where to make a road to bring up our machinery got back 11 A M & stayed in the house the rest of the day had the cook make some bread for the first time.

12月12日

用過米飯，早上7點出發返台灣府。上午11點，再吃些米飯。今天走了55里路，下午5點抵達葫蘆墩（Hoo-lotong, 豐原）。吃晚飯時，〔樂〕協台堅持要我吃米飯須學他們，伴著豬肉與糖下肚；但我可不吃這一套，反過來要他先示範吃糖配米、肉。他照作了，氣氛顯得有點尷尬。晚上阿三泡了壺茶，太濃了，嚐起來像洗碗水的味道，只好倒掉。

December 12: Started for Tia-wun-foo (Taiwanfoo) 7 A M after eating a breakfast of rice 11 A M had some more rice traveled 55 li & arrived at Hoo-lo-tong 5 P M where we had a Chinese supper the colonel insisted on my eating meat with my rice & sugar as they do so to let him know how it went I insisted on him eating sugar with his rice & meat he ate it but it went hard Ah Sun made some tea to night (tonight) so strong that we could not drink it the next will be like dish water.

❶ 本日係由出磺坑村行抵葫蘆墩，昨日並未回後壠。Sampson Kuo論文，頁101，解讀為從後壠出發，並不正確。

❷ 早餐吃飯，顯見廚師昨天試作的麵包沒成功。

❸ 有豬肉沾糖下飯這種吃法嗎？亦或是一道現今洋人喜嚐的糖醋肉（排骨）？為何兩位老美不吃豬肉呢？據絡克晚年受訪時表示，「英國領事何藍田曾警告我們注意飲食衛生，台灣水質欠佳，須煮開才能喝；豬肉不衛生，最好不吃」。由沿途老美只在11月30日的早餐吃過豬肉，其餘皆以其它肉類為主來看，似乎是受到何藍田話的影響，而非不賞老樂的臉。稍後偶爾也吃豬肉，但必然是新鮮的才吃。

12月13日

早上8點，從葫蘆墩出發，中途在〔大墩街仔，今台中市中區〕某軍官住處用午餐。被派來通知我們即將抵達的士兵摸魚去了，忘記通報消息，被老樂下令當眾鞭打〔屁股〕。下午4點抵彰化城，今天走了40里路。到老樂的衙門晚餐，吃到迄今最豐盛的中國菜。招來一位剃頭匠理髮，他想連我頭頂的毛髮一起剃掉，但我堅持那個地方可是不讓人剃光的。

December 13: Left Hoo-lo-tong 8 A M stoped (stopped) for dinner at a mandarins (mandarin's house) the solidier that was sent ahead to inform him that we were coming did not do it so the colonel bambooed him arrived at Changhwa 4 P M after traveling 40 li took super with the colonel at hia yamah the best Chinese supper that I have had was shaved by a Chinese barber who wanted to shave the top of my head told him that I did not shave eney (any) hair to shave in that direction.

❶ 他們上午、下午各走4小時，據此推斷出從豐原到彰化的中點恰為今台中市市中心中區，昔稱大墩街仔。

❷ 樂文祥擺盛宴，一則彌補中午顯然吃不好的遺憾；再則似也有意化解12日堅持要絡克吃糖醋肉的尷尬。老樂一直陪著他們、照顧他們，常在日記中出現，由絡克的行筆，樂文祥雖是行伍出身的大老粗，但帶兵賞罰分明、有雅量，酒量好、喜美食、抽鴉片、睡覺酣聲如雷，與兩位老美處得極好。老樂稍後趁絡克這趟返府城，偷偷地「領養」留在油井的薯條。

❸ 12月27日他們又相互理髮，顯然不滿意剃頭師傅的功夫。

12月14日

王大官人宣稱找不到換手的苦力，我們只得與老樂仍留彰化，好讓老王過足鴉片癮。畫了幾張工寮、吊架建築圖樣，好供夏道台參考。協台（副將）請我們吃飯，擺了二十八道菜，其中二十道不是肉類、就是混雜著肉。他的〔懷〕錶故障、不能走，要我幫忙修理。但是我不會，說可託人送到府城修理，只需1塊銀元；他聞言轉身入內，取來1塊錢，我只好收下。

December 14: Stayed with the colonel to let the old mandarin smoke opium who said that they could not get eney (any) cooleys (coolies) made some drawings of the house & rig that we intend to build to show the Taoti took dinner with the colonel had 28 dishes 20 of them were meat or had meat in them The colonel wanted me to fix his watch which would not run told him to send it to Tia-wan-foo & they would fix it for one dollar he went & got the dollar so I will have to take it.

12月15日

早上8點由彰化出發，走30里，途中休息一小時進午餐。再走20里，下午5點抵達莿桐巷。晚餐吃到熱騰騰的〔新鮮〕炸豬排、鴨肉冷盤、米飯，配上糖。客棧老闆娘給我們一隻雛雞，當明晨的早餐。囑阿三上街買糖，但他空手而回。老王、老樂將在此停留1～2天，等候在山區視察的夏道台下山會合。

December 15: Left Chang kwa (Changhwa) 8 A M went 30 li & stoped (stopped) one hour for dinner and went 20 li & arrived at Ching long hang 5 P M The old mandarin stoped with the colonel to wait for the Taotai who is up in the mountain & is expected back in a day or two to night (tonight) we had pork steak cold duck & rice & sugar for supper4 The landlady gave us a small chicken which we had cook for our breakfast Sent our boy Ah Sun out to buy some shugar (sugar) but he came back & said he could not get eney (any).

❶ 依據當日行走的里數，推估中午休息站可能是北斗或田尾。

❷ 兩位不諳「官場門道」的老美並未留下來恭候夏獻綸，隔天自行南下。

12月16日

〔星期日，〕上午8點離開莿桐巷，經過50里路，下午5點抵嘉義。仍投宿上次住的官邸，知縣〔吳鳳笙〕出差、尚未返城，只好囑廚子弄盤什錦炒飯（chowchow）。

December 16: Left Ching long hang 8 A M & arrived at Kagee 5 P M after traveling 50 li stoped (stopped) at the same house that we did the last time we was here the mandarin has not got back yet so we have our boy cook our chow chow.

12月17日

上午7點半離嘉義城，南行60里，天氣很暖和，下午5點半抵茅港尾。今天接手的轎夫，是自從府城出發以來看過最窮的苦力，每次休息總是在抽鴉片。晚上吃到新鮮的魚。

December 17: Left Kagee 7:30 A M & arrived at Ung-kang-bay 5:30 P M distance 60 li we had the poorest chair cooleys (coolies) to day (today) that we have had since we left Tiawunfoo everytime we stoped (stopped) they would smoke opium Had fresh fish for supper very warm.

註 茅港尾在荷治時期原為倒風內海南邊港口，位於急水溪、將軍溪會合處，故有新鮮漁穫。

12月18日

一早7點從茅港尾出發，走40里後，下午2點半抵台灣府。沒有任何來信，令我們大失所望。自上月底從此地出發，迄今返回原地前後共20天，來回旅行了175英里，卻一事無成，真不知這趟旅行有何意義？

December 18: Left Ung kang bay 7 A M & arrived at Taiwanfoo 2:30 P M 40 li was disappointed in not finding eney (any) mail here for us it is (There are) twenty days since we left here to go 175 miles & back & we have had that trip for accomplishing nothing what we went at all for is more than I can tell.

❶ 他們去程前後9天（11月29日～12月7日），返程7天（12月12日～18日），主要是前者有王大官人在拖時間；返程人數較少，且後半老王未隨行。

❷ 要到次年3月9日，簡時才首度接到信；絡克更慘，慢了近一個半月，直到1878年4月21日才接到家書。

❸ 簡時於12月23日寫的〈遠東函〉抱怨，離開故鄉三個半月，在上海被滯留一個多月，在福州一周，抵台後五天在府城，現在作了一趟浪費時間的旅行，又回到府城，卻仍未開始工作。

❹ 他們為何被召回府城？絡克有次詢問夏獻綸，夏答稱，「為了讓你們在府城好好過個陽曆新年」。

12月19日

往見〔美國代理副〕領事〔何藍田〕，以及〔英國長老教會〕德馬太醫生（Dr. Matthew Dickson）。德醫生上回給我們一盎司奎寧，我們這次再多買一些備用，教會醫院負責配藥的人賣我們一磅8元，但德醫生告訴他們奎寧價格已較先前便宜，於是配藥生賣我們每盎司6元。這裡有5位傳教士。

December 19: went to see the consul & Dr. Dickson who gave us an ounce quienine (quinine) when we was here before there was five mishonerys (missionaries) there who has charge of the medican (medicine) & they charged us $8.00 but the doctor said quinine was cheaper than it was so they came down to $6.00.

❶ 1877～78年，英國駐府城署領事為費里德（Alexander Frater）；約在78年底或79年初轉至擔任駐淡水領事，1884～85年清法戰爭期間表現突出。絡克指的領事係兼任美國副領事的何藍田。

❷ 5位傳教士指當時在南台的英國長老教會成員德馬太、李庥（Hugh Ritchie）、甘為霖（William Campbell）、巴克禮（Thomas Barclay）、施大闢（David Smith）。德馬太1871～76在台，1876年8月以後離台；似於1877年再度來台，待至1879年。

❸ 絡克原文的寫法不夠正確，乍看之下，會誤以為似乎是其他牧師賣藥給他；但那是不正確的，因為牧師不負責、也不會看病、配藥。而是由歷任傳道醫生訓練出的配藥生負責，如最早跟隨老馬雅各醫生（James L. Maxwell）的黃嘉智、高耀皆為配藥生出身，後來都自行開業。筆者據此，加予改寫。

12月20日

清晨7點，道台遣人來告，他因爲生病，今天不克接見我們，明天再會面。下午，沿城牆上面走了兩個半小時，繞行一周約6英里，俯瞰整座府城。與簡時打賭「1876年費城博覽會閉幕日，葛蘭特總統並未出席」，賭注一套西裝。氣溫華氏80度。

December 20: To day (Today) the Taotai was sick & sent us word that he could not see us until tomorrow at 7 A M in the afternoon took a walk around the city on the tope (top) of the wall we was two & one half hours going around about six miles Maid (made) a bet with A P Karns of a suit of clothes that U S Grant was not at Phile on the closing day of the exibishion (exhibition) in 1876 Therm (Therm.) 80°.

註 簡時〈遠東函〉陳述他對府城與台灣的印象：「此城人口約5萬，住在高25呎、厚12呎、周圍6哩長的城牆內，城牆夠寬、上可跑馬。氣候十分暖和，陰涼處介於華氏75度到80度之間（攝氏23.9～26.7度），人們整年都可睡在戶外。本島農產品很多，有鳳梨、橘子、柚子、香蕉、梨、甘薯、甘蔗、花生，以及煙草等。整體而言，這是一個美麗的地方。」

12月21日

拜會道台。中午道台的幕僚設宴款待，府城守將作陪。

December 21: Went to see the Taotai & took dinner with his sub mandarins &

the commandar (commander) of the fort at this place.

12月22日

上街採購食物，買到咖啡、餅乾、奶油、煉乳。回家推開房門，遭小偷了，煮熟的雞、鴨各飛掉一隻。

December 22: Went around the town to see what we could find in the shape of provision found some coffee crackers chow chow butter & condensed milk while we were away some one stole a chicken & duck that we had cooked & left on the table in one room.

12月23日

〔周日，〕到港邊釐金局（Likin Office, 原文寫爲Lankin）買到一雙皮鞋3元、一瓶雪莉酒9毛。返住處，下午2點吃午餐。飯後外出散步，返住處剃鬍鬚，但剃刀太鈍，找到一家礦油行，磨利刀片，待明早剃個乾淨。

December 23: Went down to Lankin bought a pair of shoes for $3.00 one bottle of sherry 90 cts came back & had dinner 2 P M (P.M.) went out for a walk came back & tried to shave but the razor was so dull that could do nothing found oil store & Sharpened razor will shave in the morning.

12月24日

道台送我們很多餐具，我們保留一些，其餘派人送回。萬事具備，但轎子不來，只好明天再啓程赴後壠。晚間，道台遣人來告，因作業不及關係，我們須再多待4～6天。

December 24: Taotai sent us a lot of table were (tableware) some we will keep and the rest we will send back got all ready to start for Oulan but the chairs didn't come so we put it off until tomorrow To night (Tonight) the

Taotai sent us word that he would not be ready for us to go for 4 or 6 days.

註 什麼「作業」呢？原來是老美的生活津貼尚未核下，詳次年元旦日記。

12月25日

〔聖誕節，〕整天待在住處。副官（sub mandarin）來訪，以湯、魚、鴨、米飯、蛋糕、餅乾、奶油、醃什錦菜、加糖及煉乳的咖啡招待他。今天氣溫介於華氏70～76度之間。

December 25: Stayed at home all day received a visit from the sub mandarin For dinner we had soup fish duck rice cake crackers butter chow chow coffee sugar & condensed milk The thometar (thermometer) stands from 70° to 76°.

❶ 這名「副官」不知是否為曾於同治十一年七月、光緒五年六月兩度短暫代理過台灣兵備道的台灣府營務知府周懋琦，或只是個小幕僚？待考。周最後一次代理道台，係因夏獻綸病死任上。

❷ 兩位老美的平安夜、聖誕節過得如此平凡，沒參加府城洋人圈的慶祝活動，因此未能替後人多留下一些卜居當地的洋人人名、軼事、生活情況，實在可惜。

12月26日

上午逛街，下午待在住處。2位漢人來訪，但彼此言語不通〔，只能比手畫腳、相互傻笑〕。

December 26: Went around town in the forenoon in the afternoon stayed at home received a visit from two Chinese but could not talk to them.

註 他未特別註明這次的住所，可能仍是上次住的台灣府學宿舍。

12月27日

與簡時互剪頭髮。下午散步時，看到漢人埋葬死者的喪禮。

December 27: Had Karns cut my hair & I cut his in return in the afternoon went out for a walk & see the Chinese bury one of ther (their ?) dead.

註 之後，他們可能長達七個半月未理髮。詳次年8月11日。

12月28日

整天都未外出。試做一支竹笛，但吹不出聲音。

December 28: Stayed at home all day tryed (tried) to make a fife out of bamboo but did not succeed.

12月29日

上午求見道台，他外出；只好下午再去，談妥下個月2日可安排好一切、啓程赴後壠。

December 29: Went to see the Taotai but he was not at home so we had to go again in the afternoon made arrangements to start for Oulan on the 2nd of next month.

12月30日

星期天。昨晚隔壁的漢人朋友有聚會，他們演奏樂器，吵得我們直到凌晨2點才睡著。整天待在住處〔補眠〕。用日晷儀測出正確的時間，發現我的懷錶快了4分鐘。室內溫度華氏83度。

December 30: Sunday Stayed at home all day made sun dial (sundial) to find out the right time watch was four minutes to fast Last night our Chinese friends had some kind of party which keep (kept) us awake until 2 A M with ther (their) music Theometar (thermometer) 83° in the shade.

12月31日

未外出。道台派人送來兩隻鴨、兩隻雞、四分之一隻山羊肉，好讓我們過年加菜。他還送了一座鐘、一盞油燈，供我們在後使用。室內華氏87度。

December 31: Stayed home all day the Taotai sent us two ducks two chickens & one quarter of goat for our new years (year's) dinner he also sent us a clock & a lamp to take with us to oulan Theom (Therm.) 87° in the shade.

註 對漢人，「送鐘」≒「送終」；洋人無此忌諱。

1878年1月1日

星期二。未外出，派僕人上街買瓶香檳酒，要價1.25元。道台派人送來我們兩人一個月的生活費津貼，共美金100元，扣掉官方代買物品費美金8元，實拿92元。室內華氏82度。

January 1, 1878: Thursday (Tuesday) Sent one boy out for a bottle of champagne which coast $1.25 stayed at home all day The Taotai sent us one monthly allowance of $700 ($100) after deducting $8 which he had paid out for stores for us Thormeter (Thermometer) 82° in the shaid (shade).

註 依據合約，他們的薪水與生活費金「按美國金錢價算」（dollars gold）；每人每月各津貼50洋銀（美元），因此共100美元，原文誤植為700美元。但絡克對所購買的酒、鞋、藥品等，均未特別註明是清幣或美金，似以當地通行的貨幣較有可能。

The same day & arived
in San Francisco on the
17th stoped at the
Reicat House
12th Went & see uncle
Hunnable. bought ticket
for Shanghai for $31.50
Left 12 noon on the
City of Peking
13th Engin broke down 6
in the evening did not
get started untill 2
in the morning
24 crossed the 180 degree
line jumped the 25th
Oct 3 arived at Yokohama
three Chinese died on
the way from San Francisco
4th Left Yokohama on the
Tokio Maru (old New York)
arived at Kobe on the
6th went to Kioto
7th went back to Kobe
8 Left Kobe 4 P.M.
went through the
inland sea beautiful
scenery
10th arived at Nagasaka
went on shore & looked
through the town
11 Left Nagasaka and
arived in Shanghai
on the 13th 4 P.M.
Stayed in Shanghai one
month went through
the old City to the
Chinese Water & all
the places of interest
25 sent a box home on the
ship ...

▲絡克於一八七七年十一月底在府城之日記（取自Sampson Kuo書，陳政三翻拍）

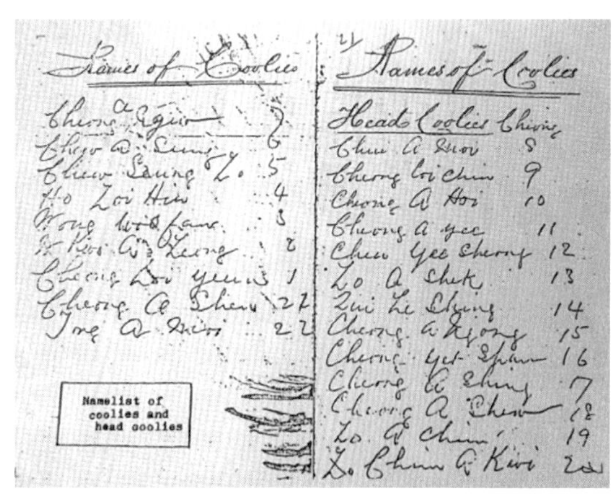
Names of Coolies
Cheong A Egin 7
Cheo A Sung 6
Chew Sung Z. 5
Ho Loi Hin 4
Wong Wid Fane 3
A Kioi A Leong 8
Cheong Loi Yeun 1
Cheong A Shew 27
Ing A Moi 22

Names of Coolies
Head Coolies Cheong
Chew A Moi 8
Cheong Loi Chin 9
Chong A Hoi 10
Cheong A Yee 11
Chew Yee Cheong 12
Lo A Chek 13
Lui Le Ching 14
Cheong A Kong 15
Cheong Yet Shun 16
Cheong A Shing 17
Cheong A Shew 18
Lo A Chin 19
Lo Chin A Kiei 20

Namelist of coolies and head coolies

▲絡克持有之公頭及苦力名單

▲出礦坑一〇六號井（陳列館展示照片，陳政三翻拍）

二部曲

等待機器・等待挫折

上山下海的日子

1878年1月2日，簡時、絡克兩位美國鑽油技師再度揮別府城，還是沿著那條簡時形容的「福爾摩沙無道路（roads），只有步行小徑（foot paths）」北上。王姓老官吏仍然陪著，但到了嘉義城的隔天清晨，天空下著毛毛雨，前晚揚言「假如下雨，大家關門睡覺」的老王果然沒出現。無奈，只好自行出發，7日到了彰化苦等了一天，失蹤三天的老王這才姍姍來遲，當晚氣溫華氏53度（約攝氏11.7度），爲日記中該年冬天台灣最低溫的記載。

1月8日上午，正待出發，可那王大官人老毛病復發，「說他今天無法同行」，只好讓他留在協台府邸過足煙癮。副將（協台）樂文祥派遣9名士兵沿途保護美國技師的安全，「阿兵哥扛著堪用的馬士奇火槍（muskets），但卻未攜帶發射彈丸用的添裝火藥；他們整天東晃西走，有一半的時間，沒看到任何士兵護衛在我們附近」，絡克抱怨道。當晚夜宿葫蘆墩（豐原）。9日一早趕路，途中很少休息，傍晚終於抵達一處不明小村，根據*Mackay's Diaries*（《馬偕日記英文版》），該地名爲Khoe-chiu（溪洲），也就是往後七個半月的住處。小村在後龍溪上游溪北，距溪南的開礦村出磺坑油井約5華里（1.79英里），四周遍植竹林、只有兩個出入口，村民約1,000人，屋子用泥磚砌牆、稻草蓋頂構成，晚上有更夫敲打更鼓守夜、報時。村中廟宇的屋頂用瓦片覆蓋，當中一間爲供神的地方，兩側偏間之一兼做村塾，兩位鑽油工匠合住在小廟的另一偏間，一住就待了在台的大部分時間（1878年1月9日～8月22日），絡克稱它為「家」（home）。這個地方，根據距離、方向，可能在公館鄉福基村內。而「溪洲」，顧名思義，可為溪畔任何一處沙洲地，並非後龍鎮的溪洲。

回到家，一切安好，只除了在烏坵島領養的那隻叫「薯條」的狗狗不見了，原來被老樂到此處視察時，順便帶回彰化豢養。抵達隔天，他們想到油泉看看，陪他們來的部分士兵，也就是絡克晚年回憶所稱的「保鑣」（bodyguards or personal soldiers），卻害怕山上的「印地安人」（Indians）而不敢伴隨上山。絡克因此試射了他那把左輪手槍，一則自己壯膽打氣，再則向不知有否躲在森林深處偷瞥的土著示威。後來唐景星送他們一把美製雷明頓步槍（Remington rifle）防身，但都未派上用場。絡克都使用「印地安人」稱呼台灣原住民，顯然將原住民比擬成美國印地安人。日記只在1月19日出現「生番」（savages）是唯一的例外，因為那天從油泉返家路上，遇到兩位漢人扛著一具被馘首的無頭屍下山。之後，3月間又陸續傳出多起印地安人（原住民）出草的消息，到3月底，已經有6位被馘首的無頭屍，另發生一起被害者雖保住腦袋，但終傷重死亡的意外。官方因此懸賞每捕殺一位山區土著，可獲50銀元的重賞，連幫他們製造大牛車的張工頭也趁著造車空檔，帶工人上山打印地安人去了。

簡時、絡克甫抵油村，立即尋覓動力來源——煤，不過遍找未獲。雖然找到油井旁噴出的瓦斯，初步認為足夠供噴氣鍋爐運轉之用，但似乎未解決如何用管線接通的問題。所以後來只能偶爾使用木材或遠從基隆運來的煤炭。1月17日，唐景星首度來到油村，還帶來一位譯員，唐上到工人不敢砍伐的神木處，一番求神問卜，假託神意云可砍，迷信的工人這才釋懷。17日當天，接到故鄉寄來的報紙；2月13日，收到布郎（R. M. Brown）轉來兩份故鄉提塔斯維爾報紙、一份美國科學雜誌。家書呢？直到3月9日，簡時一口氣接到七封信，絡克卻一封也沒收到，要等到4月21日，離鄉七個半月，他才收到家書，而且一口氣來了十四封。

爲解決機器從後壠到油井的運送問題，必須築路、建造特大號牛車。後者由絡克設計，用來裝載笨重的兩具鍋爐，須使用三頭牛拖拉。由於絡克未對這輛大牛車外型有任何著墨，只能推測它「疑似」美國西部長征篷車的改良無篷大車。本以爲可順利完成車子，但終因缺乏鐵料，大牛車直到5月1日才完成，中間絡克曾一度氣沖沖地跑下山，自掏腰包購買工具、器材。築路工程也不順利，才剛要著手就碰到「過年」，漢人傳統的過年可長可短，短的到農曆正月初五「隔開」，長的可要直到正月十五元宵節之後才有人上工。而工人又大多是附近的農民，每當農忙，常缺人手。諸事不順，加上生活費未按時發放、翻譯不夠盡責、官方後勤支援欠佳等因素，3月底絡克再也按捺不住，寫封抱怨信給人在基隆的唐景星；實事求是、講求企業精神的的唐景星透過管道，作了一番人事調動，以及作業改善，包括鴉片鬼老王調離、新管理官吏派任、增強兵力、新譯員的派任等。

等待船舶、等待機器的「看海的日子」很難熬，2月間，體貼的廚師、僕人上演一場「爭寵風暴」，讓簡時、絡克不至於太無聊。日記編號第一號的僕人阿三與第二號的阿土可能發生不快，後者出走，終被開除；新聘的阿溪才剛就任，又與阿三幹了一架，仍是前者出走。備受老美寵信、能說一口洋涇濱（pidgin English）的阿三總是贏家，獲得簡時、絡克授予「個體戶」頭銜，可決定另一助手人選。令老美訝異的是兩天後，阿三居然又把阿溪找了回來，當然，阿三現在便是阿溪的「老闆」，因此暫時解決「僕人間的戰爭」。

無聊、苦悶總得排遣，某天晚上找老士官長到公館街聽戲去，當然其他人用聽的，美國佬則用看的。只不知台上唱的是客語、福佬語、亦或京劇戲？幸好偶有好奇的、或體貼溫柔的軍士官兵眷屬來訪，總算熬過這段充滿挫折的日子。載運機器的船舶終於來到，可是第一次的喜訊卻是「狼來了」！原來，常出狀況的王姓老官僚，可能獲報機器已在淡水，想當然爾地認為依照航程、不到半天就可運達，因此「凸槌」，誤發情報，讓老美喜孜孜地跑下山、又「齪幹辣譙」的回到山上。幸好第二次的「船來了」不是「狼來了」，3月15日，第一批機器抵港；4月3日，第二批機器入港，但因風浪太大，只卸下少許，就被迫折返淡水；還好，4月9日所有剩餘的機器終於全部到港。當然還得等候大牛車、道路的完工。

兩位美籍技師利用這段空檔，搭建油泉的井架、鑽機、遊樑、基石、橫木，同時不忘派遣士兵到基隆港帶回從香港訂購的麵粉。曾搞砸至少兩次麵包，還思長進的廚師總算發憤圖強，不再烤壞麵包，香噴噴的土司夾豬肉除外的任何肉類，終於呈現在感激涕零的異鄉客桌前，物質生活暫時獲得滿足。精神苦悶的日子，也因馬偕牧師的到訪獲得紓解，因此而有機會參加獅潭底（苗栗縣獅潭鄉）教堂的禮拜儀式。令他們驚訝的是，新到任、任事積極的翻譯小陳，居然主動地催他們趕緊把儲存在港口的機器搬運上山，事情似乎有了轉機……

▲後龍城外基督長老教會內掛的馬偕像（陳政三攝）

1878年（光緒四年）

1月2日（光緒三年十一月底）

上午9點，微風細雨中，從台灣府啓程赴後壠（按應是溪洲庄），下午4點30分抵達茅港尾（下營茅港村）。晚上雨勢變大、雷電交加。前趟與我們同行的官員老王，此刻正躺在隔壁房間「快活似神仙」呢。

January 2: Left Taiwanfoo 9 A M for Oulan arrived at ung kung bay 4:30 P M Light rain this morning to night it is raining quit (quite) hard with thunder & lightning The old mandarin that was with us before is with us now at present he is in the adjoining room smoking opium.

1月3日

星期四。今天持續傾盆大雨，無法出門，整天都被困在客棧。這家旅館雖名爲客棧，但卻沒有鋪設地板，雞、犬、豬隻在屋內大搖大擺、來去自如，一點都不怕陌生人。王大官人整天抱著鴉片煙斗不放。氣溫約華氏60度。

January 3: Thursday: To day (Today) it rained so that we could not travel so we have been shut up in the house all day if you call it a house it has no floor & the dogs chickens & hogs have every thing there own way The old mandarin has keep (kept) his opium pipe red hot all day Theormeter (Thermometer) about 60 °.

1月4日

上午8點離茅港尾，傍晚6點抵嘉義城。支付轎夫每人400錢（400 cash），打發他們走。晚上下起雨，我們決定如果明天雨勢不大，仍繼續行程；但老王卻說假如還是下雨，大家關門睡覺，實在毫無敬業精神。

January 4: Left Ungkangbay 8 oclock (o'clock) & arrived at Kagee 6 P M payed (paid) our chair cooleys (coolies) 400 cash to get rid of them to night (tonight) it is raining but we have concluded to go to morrow (tomorrow) if possible. the (The) mandarin wants to sleap (sleep) the worst kind.

註 銅錢與海關兩的兌率每年不同，如據1882年海關資料，1海關兩可兌換2,000銅錢；92年1海關兩約為1.07美元。以此換算，400錢約為美金2.1角；與絡克晚年回憶，每名苦力每天可得美金1.8角，將匯率變動因素考慮進去，兩者相去不遠。

1月5日

星期六。清晨下著細雨，我們過了9點才出發，未看到王大官人現臉，不管了，就讓他過足煙癮吧。下午5點半抵莿桐巷，老闆娘弄了一隻雞、四顆蛋給我們晚餐下飯。沒看到老王在莿桐巷露臉。

January 5: Saturday This morning there was a light rain so we did not get started until after 9 oclock (o'clock) we left the old man smoking opium & have not seen him since arrived at Chang long hang 5:30 P M the land lady gave us a chicken & four eggs for our supper.

1月6日

〔禮拜天，〕今晨8點出發，整天辛苦趕路，傍晚6點下起傾盆大雨，6時15分抵達彰化城。未見到老王，事實上從抵達嘉義後，就沒再看到他了。

January 6: Started this morning at 8 oclock (o'clock) traveled hard all day & arrived at Chang hwa 6:15 P M it commenced raining at 6 oclock & is raining quit (quite) hard yet have not seen eneything (anything) of the old man since we left Kagee.

1月7日

整天待在樂副將的官邸，等候在嘉義即失蹤的王大官人，傍晚5點，他才姍姍來遲。下午觀看漢人宰豬，屠夫把豬綁在長凳上，用刀〔從脖子〕刺入，就如我們的宰法一樣；不同的是他們用盆子承接流出的豬血，然後在垂死的豬面前焚燒金紙、燃放鞭炮，豬頭鋪蓋六張灑上豬血的紙錢，豬身內外各再放置三張金紙。氣溫華氏53度（約攝氏11.7度）。

January 7: Stayed at the colonels (colonel's or colonel's house) all day waiting for the old man who we left at Kagee He came in at 5 oclock (o'clock) this evening This afternoon the Chinese buched (butchered) a hog they layed (laid) the hog on a bunch (bench) & then stuck him the same as we do but caught his blood in a basen (basin) while he was dieing (dying) they burnt joss paper in front of him & fired of (off) fire crackers they spred (spread) six sheets of paper in front of him & sprinkeled (sprinkled) blood on them & then placed three on the other in side & three outside where they burnt joss paper theom (therm.) 53°.

註 本日為該年冬天台灣最低溫的記載。

1月8日

上午9點離開彰化城，老王說他今天無法同行，只好讓他留在協台府邸繼續「養精蓄銳」。樂協台派遣9名士兵沿途保護我們的安全，士兵扛著堪用的馬士奇火槍（muskets），但卻未攜帶發射彈丸用的添裝火藥；

他們整天東晃西走，有一半的時間，沒看到任何士兵護衛在我們附近。天氣極冷、北風刺骨，下午4點半抵葫蘆墩（豐原）。

January 8: Left Changhhwa 9 A M & arrived at Hoo lo tong 4:30 P M The old man said he could not go to day (today) so we had to leave him at the colonels The colonel sent nine soldiers with us to see us safe through They are armed with good muskets but none of them has eney (any) powder to load them with they go where they please & half of the time we don't know where eney of them is It has been quit (quite) cold to day strong north wind.

註 Musket為前裝彈丸、後添火藥的舊式槍械，火繩槍（matchlock）或燧發槍（flintlock）均屬之。

1月9日

早上8點半從葫蘆墩出發。苦力行進十分緩慢，爲趕時間，所以中午沒有休息進餐，沿途也很少打尖。走到離目的地（溪洲庄，今苗栗縣公館鄉福德村內小聚落）不到5里地，苦力拒絕再前進，無奈，只好臨時多僱3名行李挑夫。我們下轎渡河（按後龍溪上流），弄得下半身濕漉漉的，傍晚6點半抵家（home），一切與離開前一模一樣，只除了狗狗不見了，被老樂到此處視察搭蓋工寮進度時，順便帶回彰化豢養。

January 9: left Hoo li tong 8:30 A M (A.M.) the cooleys (coolies) went so slow that we didnt (didn't) stop for dinner walked most of the way when we got within five li of this place the cooleys (coolies) would not go eney (any) further so we had to hire three more to carry our baggage & we walked in crossing the river go my feet wet got home 6:30 P M (P.M.) & found everything all right except the dog who went home with the colonel who was here to see some men at work building our house.

❶ 根據*Mackay's Diaries*（p. 241）、《馬偕日記》（玉山社，頁333），絡克指的「家」位於後龍溪上流北岸公館鄉溪洲小村（應在溪北福德村內），是座廟宇。工寮在1.79哩（2.88公里）外、溪南出礦坑油泉旁。Sampson Kuo在他的博士論文中（頁105），解讀為回到後壠，並不正確。

❷ 由於「薯條」被樂文祥帶走，失去這條狗，似乎使得異鄉客少了排解寂寞的管道；真不知老美為何未將薯條要回？「烏坵狗過台灣」的記載不再出現於日記。而不論薯條是公是母，牠的後代顯然應該已遍佈彰化地區；除非喜愛美食的老樂另有盤算，他該不會把牠當成「熱狗」吧？

1月10日

〔士兵今天才出現，〕他們一到，我們立即準備出發到油泉湧出地（oil springs），前往視察工人搭蓋工寮的情形，士兵懼怕「印地安人」（Indians），死也不願上山，只有3名肯陪我們；於是我們與阿土（Ah Law）、三兵前去，但沒看到任何原住民。

January 10: As soon as the soldiers came we go (got) ready to go up to the oil springs where there is (were) men building our house the soldiers all backed out but three they was (were) afraid of the indines (Indians) so we took Ah Law & the three soldiers & went but did not see eney (any) Indines.

❶ 絡克大都使用indins或indines（Indians）稱呼台灣原住民，只在1月19日出現savages是唯一的例外，他顯然將原住民比擬成美國印地安人；尚稱妥當的比擬。

❷ 因原稿字跡潦草，Ah Law可能與2月18日的Ah Tow是同一人，也就是他們的廚師、第二僕人。

1月11日

到上游糖廍觀看漢人製糖。四處搜尋煤炭，但沒找到。第一次試射我的左輪手槍（revolver）。氣溫華氏55度。

January 11: Went up to the shugar (sugar) works (workshop) & see the Chinese make (making) shugar looked around to see if we could find eney (any) coal but could not find eney shot my revolver for the first time Theom (Therm.) 55°.

❶ 糖廍可能在大湖鄉內，後龍溪上游、大湖溪畔的蔗部坪（蔗部坑）。

❷ 他後來又擁有一把步槍，詳8月11日。

1月12日

星期六。據說下游有煤炭，〔早上〕順河走了7里路，只發現不多的木柴。下午到上游油泉視察工寮進度，約20名左右的苦力正在製作磚塊、刨平地面。回到住處，赫然發現王大官人正等候我們。

January 12: Saturday Went down the river 7 li to see some coal that we heard that was down there but it turned out to be wood & not much of it at that in the afternoon went up to the oil springs where there is (were) about 20 cooleys (coolies) making brick (bricks) & grading for a house came back and found the old mandarin waiting for us.

1月13日

〔星期天，〕深入上游山區，直到走不通爲止。沿著一條支流小溪試圖下山，但有多處塌陷，無法走通，差點回不了家。

January 13: went up on the mountain as far as we could go had some trouble getting back as we tryed (tried) to follow a small stream & there was (were) so many falls that we could not get around.

1月14日

帶7名兵上到油泉，再從那裡爬上山區，找到一棵夠大的樹；士兵稱更上面近山頂處，還有一些大樹，但他們害怕「印地安人」，不肯陪我們前往。我們上到靠山頂處，果然找到那些大樹。在油泉後方海拔約2,500英尺（762公尺）靠近山頂處，發現那裡噴出的瓦斯量，足於轉動蒸氣鍋爐。

January 14: went up to the oil springs with seven soldiers from there went up on the mountain to see some timber(s) found one tree the soldiers said there was (were) some more up farther but they was (were) afraid of the indins (Indians) & would not go found a place near the top of a mountain back of (to) the oil springs about 2500 feet & 2500 feet high where the gas came up enough to run a boiler.

1月15日

上午到油泉，釘好要架設鑽機（rig）的木樁，返住處用午餐。下午在家休息，簡時（A. Port Karns）弄好麵團準備做麵包，但卻被廚子給烤焦了！

January 15: went up to the oil springs staked out a spot to build a rig came back had tiffin & stayed in the house all afternoon Karns made some bread for supper but the cook spoilt them in backing (baking).

❶ 這是廚師第二次搞砸他們心愛的麵包；第一次在去年12月11日。之後未再提及，應是廚子已學會烤麵包的技術。

❷ 簡時〈遠東函〉寫道，「出磺坑附近有四處油泉，幾世紀來，在地人每天可撈取4～5加侖漂浮在水面的石油」。此處是他們選擇的第一處鑽鑿油井的地方。

1月16日

上到油泉。漢人云不敢砍那棵我們看過的神木，因爲他們深信砍神木的人會生病、然後死亡。我們告訴工人先放過那棵樹，另找其它適當的樹木。

January 16: Went up to the oil springs The Chinese said that thay (they) could not cut the tree that we went to see Thay say that the man who cuts it down will get sick & die so we told them to go & find other trees & let it stand.

1月17日

上午到油泉，中午返家吃午餐。下午〔唐〕景星（Dick Sine）抵達這裡，他全權負責鑿油事務，還帶來一位譯員供我們使喚。收到故鄉提塔斯維爾（Titusville）寄來的一份報紙。

January 17: Went up to the oil springs came back for tiffin Dick Sine arrived this afternoon he has full charge of the oil business now he brought an interpreter with him to stay with us received mail from Titusville one nasepaper (newapaper).

1月18日

陪同尊貴的景星到油泉，請他上到那棵神木處看如何解決，他祈求神明、詢問是否可砍那棵樹？神說可以，工人也就釋懷了。

January 18: Went up to the oil springs with his Excelinacy (Excellency) Dick had him go up to the large tree & ask the joss if we could cut it they said we could so that ia all right.

1月19日

星期六。早上到油泉，返程路上遇到兩位漢人扛著一具無頭屍下山，是在6～8華里外被「生番」（savages）馘首的。下午去見道台〔唐景星〕，他囑我們做一個載運鑽油機械的車子（wagon）模型，好帶到後找人複製。晚飯後開始做模型，今晚完成一半。

January 19: Saturday Went up to the oil springs when we was coming back we see two Chinese carrying the body of a Chineman (Chinaman) that had been beheaded by the savages about six or eight li from here. In the afternoon went to see Taotai who wanted us to make a modle (model) for a wagon for him to take to Oulan to get one made by it for us to bring our machinery up on went 〔to〕 work on the wagon after supper I got it about half finished.

❶ 本日是絡克日記唯一使用savages的一次，其餘都用Indins。

❷ 唐景星擁有「觀察」虛銜，約略等同「道台」，〈1878年英國淡水領事報告〉稱其擁有道台銜。絡克在油井附近提到的道台，都指唐，只有1878年4月17日例外。美國技師在台期間，根據絡克的日記記載，夏獻綸似未曾到過油井視察。

1月20日

禮拜日。完成貨車模型，下山將模型交給唐道台，唐景星將模型交予一位自稱能照設計樣式製作的木匠，只要索價不離譜，就讓他承包。道台向我們道歉，說他忘記今天是星期日，還要我們趕工。愛抽鴉片的王大官人今晚來訪。

January 20: Sunday Finished the modle (model) for the wagon and went down to the taotais (taotai's) with it he sent out for a carpenter who said he could make one so the taotai told him to give him his price & if it was not high he could make it. The taotai said he forgot that to day (today) was sunday & that we would have to exclose (excuse) him Mr. Wong the opium smoking old mandarin (mandarin) call on us this evening.

❶ 後龍（壠）距離溪洲庄30里，以當時每天旅行可走50～60里來估計，唐景星顯然不是住在後龍，否則老美無法當天來回；駐紮後龍的老王也不須上山造訪他們。唐可能住在公館或福基，也就是木匠住的小村；而以住公館可能性較大。詳1月25日。

❷ 木匠可能是本年3月22日提到的張領班，他造的是一輛美式大型牛車，足足花了三個多月時間，5月1日才完成。

1月21日

僱8名工人（coolies）搭建橫跨〔後龍〕溪的幾座步行便橋（foot bridges），以便從住處小村通到油泉，每人每天支付750錢；造橋工程費時，可有他們忙的。

January 21: Had eight cooleys (coolies) building foot bridges across the creek between here & the oil springs The eight cooleys received $750 per day & thay (they) do about that amount of work.

註 $750應為750錢（cash）；這批工人顯然不是普通苦力（每天領美金1.8角），而是有技術的工人，所以每天領取約美金3.9角。兌率詳1月4日註釋。

1月22日

早上8點準備出發到油泉，但造橋的苦力在沒拿到工錢之前不願上工。下午監督〔已拿到工錢的〕工人建橋，並帶4位士兵、4名苦力到上游8里許處，砍取搭橋用的竹子。

January 22: was ready to start for the oil springs at 8 oclock (o'clock) but the cooleys (coolies) would not go until thay (they) received ther (their) pay so we did not go until this afternoon worked on foot bridge & went with 4 soldiers & 4 cooleys up the river about 8 li after bamboo poles to put on the bridge.

1月23日

上午下雨，未上工。下午苦力說太冷，所以我們整天就窩在屋內。今晚溫度華氏62度（約攝氏16.7度）。

January 23: This morning it rained so we did not go to work & this afternoon it was to (too) cold for the cooleys (coolies) so we stayed in the house all day Theom (Therm.) to night (tonight) stands at 62°.

1月24日

去見唐道台，商量從此處修造一條通往後壠的道路事宜，最後，決定明天開始，由我們在擬建的路線先打上木樁，好讓苦力據此開始修路。前往視察便橋、工寮的進度，苦力正趕工中。

January 24: Went to see the Taotai about building a road from here to Oulan concluded to go to morrow (tomorrow) & stake out the road so the cooleys (coolies) can go to work on it Went up to see house the cooleys were getting along with & the bridges Light rain this morning.

1月25日

下山到後壠，準備打上修路木樁；唐道台與我們同行。途中在貓里（Marlee, 1886年改名苗栗）午餐，下午5點抵後壠。沿途發現要修的路況比原先預期的來得好。

January 25: Walked down to Oulan to stake out a road the Taotai went with us

Took dinner at Marlee arrived at Oulan 5 oclock (o'clock) found a much better road than we expected to find.

註 據絡克回憶，「造車木匠住在後壠到油泉途中，二、三小鎮其中的一處」。唐與老美中途在苗栗用餐，顯見木匠、唐景星都住離溪洲油村較近的公館。但為何不是福基村呢？福基緊鄰石圍牆東北邊，離溪洲太近，步行僅約半小時，而且又是極偏僻的野村，唐實在沒理由住該地。木匠則不排除有可能住在介於公館、溪洲庄中間的福基，但仍以公館較可能。

1月26日

等候道台梳洗、用餐完畢，上午9點出發到〔後壠〕港口，選定一處適合卸下機械的地點，下午1點回到後壠，返程途中在貓里用餐，5點回到住處。

January 26: After waiting until 9 oclock (o'clock) for the Taotai to get ready went down to the harbor found a place to unload our machinery & got back to Oulan 1 oclock walked to Marley for dinner got home 5 oclock.

1月27日

星期天。上油泉視察工寮進度，但工程緩慢，看不出有任何進展。返住處用午餐，下午未再出門。

January 27: Sunday Went up to the oil springs to see how the men were getting along with the house but could not see as thay (they) had done anything came back to tiffin & stayed in the house the rest of the day.

1月28日

與唐道台到油泉，再爬山上到神木處，工人已將神木砍倒、正在鋸短。回家吃午餐，飯後下山視察築路情形，但沒看到有人在工作。

January 28: Went up to the oil springs with the Taotai went up to the big tree & found that thay (they) had cut it down & was (were) sawing it up came back for tiffin & went down to see if there was eney (any) one working on the road but could not see anyone.

1月29日

到油泉視察鑽機架設準備進度，遊樑（walking beam）、主基石（main sill），以及小橫木已大致備妥。唐道台今天離開、返回〔基隆八斗子〕煤港（coal harbour）。氣溫華氏75度，是近幾日最暖和的天氣。

January 29: Went up to the oil springs to see about getting out the rig the walking beam is out & the main sill and some of the small timber The Taotai went back to cool (coal) harboour to day. Theom (Therm.) 75° the warmest day we have had for some time.

註 這是唐景星首次視察油泉，前後13天。第二次在本年7月18日，想親眼看到鑽出的「第一滴油」。

1月30日

今天天氣寒冷，還下著細雨。因爲農曆新年將屆，苦力都未上工，說是要準備、採買年節用的東西。有位新到的官員在工地四處晃盪。

January 30: To day (Today) is just cool with light rain the cooleys (coolies) have all quit work on account of new year Thay (They) want some time to buy things for new years There was a new mandarin around to day.

註 工人都是附近村莊的農民，當苦力只是農閒打工性質，很注重一年一度的過年，依傳統，年節要休到農曆正月十五元宵節，因此部分人休到2月16日，17日才再上工；部分人在正月初五「隔開」之後的第一天（2月7日、農曆初六）開始工作。

1月31日

今晨，小村舉行一場喪禮。死者兩天前才過世，但新年在即，爲免破壞歡樂氣氛，所以家屬〔依據習俗〕提早將他下葬。約有20名身穿麻布衣、哀嚎得很大聲的送葬者，他們焚燒很多紙錢、香柱，香案擺著全羊、全豬各一隻，還擺上各種煮肉、醃甜肉等祭品。天氣陰雨，華氏64度。

January 31: This morning we had a Chinese funeral the man died two days ago but owning to new years thay (they) the funeral so they could have a good time There was about 20 morners (mourners) dressed in sackcloth & the noise that maid was fritfull The sacrifice consisted of one goat one hog & a lot of cooked meats & sweet meats they burnt about a bushel of joss paper & a lot of joss sticks wether (weather) clawedy (cloudy) with light rain Theom (Therm.) 64°.

註 附近村莊居民為客家人。

2月1日

〔農曆除夕，〕今天對漢人而言，是個充滿活力的日子；對看熱鬧的我們也是。他們爭相到廟裡拜拜，整天都有人帶牲品、金紙來燒香拜佛，還燃放鞭炮；要是他們持續更久一點，說不定會用噪音、煙霧把我們趕出〔所借住的〕廟（Temple）。昨天他們說今天是過年，今天又說明天是過年，在這個國家，沒有任何事情是確定的。

February 1: To day (Today) there has been a lively time with the Chinese & thay (they) make it lively for us. They have been offering sacrafices burning joss paper & sticks & firing of (off) fire crackers all day thay call it worshipping if thay keep it up much longer thay will drive us out of the Temple with the noise & smook (smoke) Yesterday they said that to day was new years (new year) To day thay said that to morrow (tomorrow) is new year nothing certain in this country.

❶ 此處透露他們住在廟宇。絡克晚年受訪回憶，那個他不會唸的（溪洲）小村距油井約1英里（日記寫為5華里，約1.79英里、約3.1公里），四周遍植竹林、只有兩個出入口，村民約1,000人，屋子用泥磚砌牆、稻草蓋頂構成，晚上有更夫敲更鼓守夜、報時。村中廟宇的屋頂用瓦片覆蓋，當中為供神的地方，兩側偏間之一兼做村塾，他與簡時合住在小廟的另一偏間，一住就待了在台的大部分時間（1878年1月9日～8月22日）。他們開採的溪南出磺坑油井處有座「油礦開採陳列館」，值得一遊。

▲ 出磺坑（公館鄉開礦村）台灣油礦開採陳列館（陳政三攝）

❷ 「過年」對華人是指新春期間；但也有可能是翻譯不夠清楚造成。

（1878年）2月2日（光緒四年元月一日）

今天是漢人的新年（new years），還有5天也是新年，村民忙著拜拜、賭博，幸好沒放太多的鞭炮，眞是謝天謝地。氣溫華氏65度，太陽在白天露過臉。

February 2: This is new years (year) with the Chinese & will be new years for five days more Thay have got through worshipping & gone to gambling not many fire crackers to day (today) & we are very thankful Theom (therm.) 65° sun shone part of the day.

註 他用複數的new years，顯然被那麼多天的「過年」給搞糊塗了。正月初五

「隔開」，代表隔天該工作了；更傳統的要過到正月十五元宵節。

2月3日

星期日。整天待在家裡。漢人持續過年期間的賭博，並進行驅逐惡魔凶鬼儀式。華氏73度，晚間下起細雨。

February 3: Sunday New years continued gambling also stayed at home all day The Chinese scared all the evil spirits out of the town Thom (Therm.) 73° light rain this evening.

2月4日

下山視察道路，回家吃午飯，然後一直待在屋內未出門。華氏65度，毛毛雨。

February 4: Went down to see the road came back for tiffin & stayed in the house the rest of the day Thom 65° Light rain.

2月5日

幾乎整天都在下雨，待在家裡。十分寒冷，我們生爐火取暖。

February 5: Stayed at home all day rained most of the day and quit (quite) cool So we had to have a fire in.

2月6日

整天待在〔廟內〕屋裡。今天又冷又濕，天氣欠佳，苦力不願上工。華氏55度，下了些雨。

February 6: Stayed in the house all day the cooleys (coolies) would not go to work to day (today) on account of the wether (weather) which is wet & cold (55°) with some rain.

2月7日

築路工人今天上工，我們下山去看築路情形，順便〔到公館〕探視用來載機器的大牛車製造進度。王大官人〔從府城過完年〕回來。接到道台〔唐景星〕的信，內云機械已經運抵福州，將儘快運來。

February 7: The cooleys (coolies) went to work to day (today) on the road so we went down to see how thay was getting along went & see the wagon also Mr. Wong came back to day Received a letter from the Taotai stating that out (our) machinery was at Foochow & would be sent on as soon as possible.

2月8日

想去油泉，但其中那座已搭好的便橋被高漲的溪水沖走，無法過河，只得折回。接獲唐道台來信，內稱機械已運到淡水（Tomsuey, 之後也用過Tomsuia, Tamsui）。

February 8: Started to go up to the oil springs but the water has (had) carried of (off) one of our bridges so we could not get accrossed the creek Received letter from the Taotai stating that our machinery was in Tomsuey.

註 此時唐景星人在北部，或至少在八斗子煤港有輪船招商局辦事處，因此可確定7、8兩日提到的道台是他。

2月9日

今晨接到〔人在後壠的〕老王來信，宣稱機器已運抵後壠；我們興沖沖地趕去，想看卸貨情形，抵達才知機械尚未運來，而且也沒人知道何時會到。

February 9: This morning we received a letter from Mr. Wong stating that the machinery was at Oulan & that we was to go & see to unloading it came to Oulan & found that our machinery had not arrived & thay (they) did not know when it would.

❶ 常出狀況的王姓老官僚，可能也獲報機器已在淡水，想當然爾地認為依照航程、不到半天就可運達，因此演出「狼來了」戲碼。

❷ 第一批機器在3月15日才運到。

2月10日

〔星期天，〕早上8點從後壠走到苗栗，在苗栗僱轎，沿途順道視察道路修築情形，約有130名苦力在築路。晚上氣溫華氏75度、微雨。

February 10: Let (Left) Oulan 8 A M walked to Marlee & hired chairs for the rest of the way There is (are) about 130 coolies working on the road between here & Oulan Theom (Therm.) to night (tonight) 75° some rain.

2月11日

禮拜一。到油泉，安排22名苦力修平鑿井處的坡度（grade, 原文用graid）。這個工地迄昨天，已約兩週無人上工。晚上下雨。

February 11: Monday Went up to the oil springs there has been no work done there since we was (were) up some two weeks ago Set 22 coolies to work on the graid (grade) for the well To night (Tonight) it is raining.

2月12日

下山〔到公館〕看大牛車的進展，發現有6名木匠正「慢工出細活」的在工作，實在太慢了，還看不出甚麼碗糕。返村吃午餐，未再出門。微雨、氣溫華氏60度。

February 12: Went down to see how thay (they) were getting along with the wagon found six men at work on it but they are (were) so slow that I could not see that thay had done much came back for tiffin & stayed in the house the rest of the day Light rain Thom (Therm.) 60°.

2月13日

收到布郎（R. M. Brown）寄來兩份故鄉提塔斯維爾報紙、一份美國科學雜誌。下山視察牛車、道路進度，實在超慢。派人送信〔到後壠〕給老王，要他趕緊付我們這個月的生活費津貼；我們錢花光了，無法買生活必需品，簡直是坐吃山空。陰天、氣溫華氏62度。

February 13: Received two Titusville papers & one scientific American from R M Brown Went down to see the wagon & the road get (got) along very slow Sent a letter to Mr. Wong to have him pay us one monthly alounce (allowance) out of money & eneey thing (anything) cloudy Thom (Therm.) 62°.

❶ 布郎是中文合約上使用的名字。

❷ 他們1月1日領到一月份的津貼；2月快過一半，錢尚未核下，效率欠佳。

2月14日

上油井看木柴準備情形，返程遭雨淋成落湯雞。回家未再出門。

February 14: Went up to the oil springs to see the timber got caught in the rain coming home & got quit (quite) wet stayed in the house the rest of the day.

2月15日

下山視察道路、牛車進度；回家吃午飯，之後未再外出。今天〔農曆正月十四〕村民舉行祭禮，晚上放許多鞭炮、還敲打銅鑼。

February 15: Went down to see the road & wagon came back to tiffin stayed in the house the rest of the day To day (Today) the Chinese had a feast & to night (tonight) they are firing a good many fire crackers & beating gongs.

2月16日

下山視察道路、大牛車進度，發現竟然無人上工！一問才知原來今天是過年的最後一天〔元宵節〕，工人須準備拜拜。牛車看起來幾近完成。晚上到處有人放鞭炮。

February 16: Went down to see the road & wagon no man at work on the road to day (today) Thay (They) have to worship as this the last day of there (their) new year the wagon is almost finished plenty of fire crackers to night (tonight).

❶ 本日又稱上元節，為天官大帝與臨水夫人生辰。

❷ 日記中，過年過節未見他們受邀到村民家吃拜拜，顯見彼此互動不多、關係不是挺好的。如4月25日，有人邀宴，也遭他們婉謝。

2月17日

禮拜日，早上微雨。收到這個月每人50美元的生活津貼（board money），派人送回上次我們兩人預支的（back money）50元。下山視察道路，有26名工人在築路。

February 17: Sunday Light rain this morning Received $50.00 for board money sent $50.00 back as back money Went down to see the road 26 men working to day (today).

註 此處語意不清，「收到五十、還五十」，不通，似為「發下一百，留下五十，還五十」之意，乃譯為每人領50，另歸還共同預支的50元；但為何不在後壠就直接扣掉？這個back money該不會是回扣吧？參閱3月7日、3月24日。

2月18日

下山看牛車，全部完成了，只剩下鐵工部分尚待安裝。已一個星期沒看到〔第二號僕人〕阿土（Ah Tow），要〔一號僕人〕阿三（Ah Sun）去把他找回了來，付他一個月的工資，當面要他走路，阿土〔見羞轉生氣，〕拒收工資。

February 18: Went down to see the wagon all finished except the irons Had not seen Ah Tow for a week so we sent Ah Sun to bring him in Payed him for a full month & told him that we did not want him any longer he refused the money.

註 阿土可能是府城請來的廚子。

2月19日

下山看牛車，回家吃午餐，下著細雨，飯後未再外出。寫信給〔妹妹〕莎拉（Sarah）。

February 19: Went down to see the wagon came back to tiffin stayed in the house the rest of the day Light rain wrote a letter to Sarah.

2月20日

下山巡視築路，回來後未再外出。僕人間麻煩不斷，阿三與第三號僕人（boys No.3, 阿溪）打了一架；所以我們同意每月付17〔銀〕元給阿三，讓他自行聘用助手。晚上雨勢加大，華氏65度。

February 20: Went down to see the road came back & stayed in the house the rest of the day more trouble with our boys No 3 got to fighting so we agreed to give Ah Sun $17.00 per month & let him hire his own help (helper) more rain to night (tonight) theometer (thermometer) 65°.

❶ 阿三每月工資10銀元（去年11月28日載）是從府城跟來的僕人，阿溪應

是接替阿土當廚子。boys在此應解讀為僕人，因為他或他們的女兒已經大到可跟簡時發生「麻煩事」了，詳本年9月2日。

❷ 4月25日絡克在寫給妹妹的信中云：「我們有兩位僕人，一位當跑腿小弟，另一位煮飯。但我常被他們搞混，因為他們都煮飯。」

2月21日

下山〔到公館〕看牛車進度，木工已完成，正在冶鍊鐵工零件部分。返家午飯，再下山巡視道路。同時，派兩名士兵到油泉了解木材準備情形，士兵回報5、6天內可備妥。華氏65度。

February 21: Went down to see the wagon wood work finished trying to make irons came home to tiffin & went to see the road sent two soldiers up to see about the timber say thay (they) will finish in five or six days Them (therm.) 65°.

註 絡克日記記載，溪洲庄到後壠需走4～5小時，依比例推估，到公館約需不到1個半小時。

2月22日

下去巡視道路施工，今天到了45名工人，據稱3天內可望築到石圍牆（the Stone Wall, 公館鄉石牆村）。阿三居然又把〔跟他幹架的〕第三號僕人阿溪（No. 3 Ah Ked, 原文為Atheked, 顯然有誤）給請了回來！華氏65度、濃霧。

February 22: Went down to see the road about 45 men at work to day (today) thay (they) expect to finish to the stone wall in three days Ah Sun killed (hired) No 3 Atheked back again Thom (therm.) 65° very fogy (foggy).

註 石圍牆在後龍溪上游北岸邊，位於第一處鑽井油泉的西側。中影電影「源」（The Pioneers）即是簡時、絡克來台的故事，不過兩人是配角，主角是相傳乾隆年間渡台的客家籍監生吳琳芳，乾隆二十三年（1758）入墾公館；嘉慶二十二年（1817）拓墾石圍牆，曾在石隙間發現「地油」滲出，兩者探採石油的年代相差約60年；影片內容有不少錯誤。其中，認定老美住石圍牆即不正確，因為絡克說他與簡時都不會拼出所住地方的村名。另外，《源》片簡時帶太太Rosy來台、住三合院……等，都非事實，更不用提當時環境下過分強調「愛國」的拍攝手法；不過兩輪牛車、油井井架、鑽油機械等道具十分真確，也非無優點。這個故事實在值得重拍。客家電視台約兩年前重拍成連續劇，用的仍是《源》編劇張毅免費提供的劇本。

2月23日

今天是兩個星期以來天氣最晴朗的日子。下山去看大牛車進度，途中見到51名工人正加緊趕工修路，終於築到半山腰的石圍牆。阿三沒買到雞。

February 23: To day (Today) was the pleasantest day that we have had for two weeks Them (Therm.) 65 Went down to see the wagon 51 men at work on the road to day finished up the stoan wall (Stone Wall) Ah Sun could not buy any chickeneys (chickens) to day.

2月24日

周日。因雨，在家休息一天，做了一具捕鼠器，擺在屋內通水口處。華氏62度。

February 24: Sunday Stayed in the house all day on account of the rain made a trap to catch rats out of the washhole Therometer (thermometer) 62°.

2月25日

整天下著小雨，我們也一直未外出。寫信給福羅倫斯（Florence）。華氏58度。

February 25: Light rain all day stayed in the house Wrote a letter to Florence Therem (Therm.) 58°.

2月26日

順河而上，信步來到小村與油泉的半路，那裡看起來似有蘊藏煤炭的樣子。中午12點回住處，未再外出。氣溫華氏66度。

February 26: Went up the river about half way to the oil springs where there is good indication of coal got back by 12 oclock (o'clock) & stayed in the house the rest of the day. Therm (Therm.) 66°.

2月27日

下山看大牛車，黑手（blacksmith）已用完所有鐵料，仍不夠，已派人再去買1銀元的材料。天氣晴朗，氣溫66度。

February 27: Went down to the wagon the blacksmith had used all his iron but has sent a dollar for some more pleasant Therm 66°.

2月28日

整天都待在家裡。華氏60度。

February 28: Stayed at home all day Theremometer (Thermometer) 60°.

3月1日

下山看大牛車；沒買到鐵料，一切陷入停頓狀態。天氣晴朗、舒適，華氏60度。

March 1: Went down to see the wagon no iron so everything is at a standstill wether (weather) pleasant thereomometer (thermometer) 60 °.

3月2日

想上去油泉，但河水暴漲，我們無法渡河，只好折返住處閒待整天。有人來告，稱王大官人已在早上從後壠出發回台灣府。氣溫62度。

March 2: Started to go up to the oil springs but the water was so high that we could not cross stayed in the house the rest of the day Received word that Mr. Wong had started for Tiawanfoo (Taiwanfoo) this morning Th (Therm.) 62 °.

3月3日

〔周日，〕工人繼續整路；我們則整天在家休息。華氏60度。

March 3: The coolie went to work on the road Stayed in the house all day Ther (Therm.) 60 °.

註 2月23日道路已修到石圍牆；這裡修的是石圍牆到東方油泉路段。

3月4日

星期一。上到石圍牆，沒看到苦力在修路。返住處休息。「額外外委」（head soldiers, 士官長）的兩位夫人一道來拜訪我們。

March 4: Monday Went up to the stoan wall (Stone Wall) no coolies at work Stayed in the house the rest of the day Received a visit from the head soldiers wives two in number.

❶ 有關下級武官階級，詳1877年12月6日註2。Sampson Kuo認為head soldiers可能是老美的私人保鑣，參閱本年5月23日註2。

❷ head soldiers未標所有格，如是「head soldier's」，則是1名外委的2位夫人；

如為「head soldiers'」，則至少2位以上外委的夫人們。

❸ 2月23日道路已修到石圍牆；本日用「上到石圍牆」，顯然石圍牆的地勢較油村高，因此用「went up to」。

3月5日

整天下雨，不得不待在廟裡。華氏65度。

March 5: Rained all day so had to stay in the house Ther (Therm.) 65°.

3月6日

上到石圍牆，天氣糟透了，苦力沒上工，折返住處，閒閒冇代誌。華氏62度。

March 6: Went up to the stoan wall (Stone Wall) no coolies at work on account of bad wether (weather) stayed in the house the rest of the day 62°.

3月7日

下山看大牛車，〔缺鐵料〕毫無進展。收到本月50美元生活費。華氏64度。

March 7: Went down to see the wagon nothing done Recd (=Received) $50.00 on board There (Therm.) 64°.

註 老王回府城之前，總算未忘發下他們的生活津貼；但只發一半。詳本年3月24日註。

3月8日

「外委軍官」（hiring officer）率領所有士兵，伴隨我們來到約距油泉半途的〔後龍〕溪邊，〔石圍牆〕那裡有73名苦力正在修路。昨晚，〔2月24日做成的〕捕鼠器逮到一隻老鼠。今天天氣十分晴朗，是兩週以來最舒適的一天。

March 8: Went up the river about half way to the oil springs where there is (are) 73 coolies working on the road the hiring officer went with us & all the soldiers caught a rat last night to day (today) is very pleasant the first for two weeks There 68°.

註 「外委」可為正八品的千總或正九品的把總。Sampson Kuo將hiring officer解讀為低階文官。文官叫得動士兵嗎？

3月9日

與簡時到油泉巡視，有4、5名苦力在蓋工寮，但看不出與前次來時有任何進展。收到去年12月21日倫敦出版的報紙5份；簡時接獲七封從上海寄出的回信；我在上海也和他一樣同時寄出八封信件，基於不明的原因，卻未收到任何回音。看樣子，至少還要再等一個月，才可能收到信件。

March 9: Went up to the oil springsfour or five men working on the house but cant (can't or cannot) see as thay (they) have done anything since we was (were) up there before Received five London papers up to Dec (Dec.) 21st Karns rec'd answers from letters he wrote at Shanghai seven in all I wrote eight at the same time but for some reason have receiced no answer dont (don't) expect to yet as other mail for a month.

❶ 舊報紙可能是布朗寄來；信件也可能是他從上海轉來。詳本年2月13日。

❷ 絡克直到4月21日才首次接到家書，多達十四封。

3月10日

〔周日，〕上山巡視〔石圍牆〕修路進度，約有30位工人在工作。晚上，村民前來告稱，明天有拜拜，依據他們信奉的宗教，不可殺生，要求我們明天不要殺雞。

March 10: Went up to see the road about thirty men at work. To night (tonight) the Chinese requested us not to kill any chickeneys (chickens) to morrow (tomorrow) as thay (they) are going to sacrifice other gods & it is against ther (their) religion to kill any thins (Therm.) 66°.

3月11日

巡視修路。今天的祭祀，似乎只是小拜拜，全村只聽到三、四響鞭炮聲，一點鼓聲而已。氣溫華氏62度，晚間微雨。

March 11: Went up to see the road the Chinese dont (don't) seem to pay much attention to there (their) sacrifices have only fixed three or four guns & drummed (drummed) a little through the town Thereometer (Thermometer) 62° Light rain this enening.

註 本日為農曆二月初八、宋帝王生。宋帝王，俗傳為陰間十殿閻王之第三殿，掌落蒸地獄、孽鏡台地獄。

3月12日

上到油泉，土水師已著手搭築工寮牆壁。在場的士兵筆劃手勢表示，「印地安人」又殺死1人，並帶走人頭。自從我們到這裡以來，屈指一算，已有4人被山區土著殺害。華氏64度。

March 12: Went up to the oil springs commenced building the walls of the house The soldiers made us understand by signs that the indians (Indians) had shot an other (another) man & cut his head of (off) this makes four since we have been here Ther (Therm.) 64°.

3月13日

白天在油泉監工。晚上下山找幾位士官長（額外外委），一齊上戲館（Chinese theatre）看戲（sing-song），直到11點才結束。據通報，機器已運抵後壠港，我們決定明早下山。

March 13: Went up to the oil springs in the evening went down to the head soldiers to a sing-song in Chinese thestee (theatre or theater) stayed until 11 oclock (o'clock) Received word that machinery is at Oulan so we will go down in the morning.

註 以距離、地方發展情形推測，附近最可能有戲館的地方應在公館街。嘉慶年間漢人入墾公館鄉，墾首大地主在公館庄伯公廟旁建租館收租，因而得名；當地水田面積廣大，青蛙非常多，不分晝夜啼叫不停，因此又稱為「蛤仔市」。可想像，大、小租戶一年兩次定期到公館繳納地租，然後上戲館（茶館）聽戲、聊天、嗑瓜子，應是不可或缺的儀式。

3月14日

僱用的轎子沒來，只好步行出發赴後壠；譯員宣稱不能走遠路，只好讓他留下。下午1點抵達後壠，與士兵共進午餐，用筷子扒了好幾碗飯。

March 14: Started for Oulan 8 A M (A.M.) the chairs we sent for did not come so we walked The interpreter said he could not walk so we left him arrived at 1 P M (P.M.) took tiffin with the solsiers more chep sticks (chopsticks) & rice.

註 他們未到碼頭，顯見貨船當天尚未抵達。

3月15日

到〔後壠港〕港口，看到兩艘戎克帆船（junks）載來一部分機器，包括兩套蒸氣鍋爐（boilers）、3、4大箱機具。船員不肯把船灣靠碼頭，因爲深怕卸鍋爐時，萬一不愼脫落，把船砸毀。無奈，跑到海邊挖貝殼、牡蠣〔出氣〕，也無所獲。海風狂吹。

March 15: Found two junks loaded with part of our machinery the two boilers are here & three or four boxes we could not get the Chinese to bring these boats up to the wharf thay (they) was (were) that we would brake (break ?) them unloading the boilers Went down the coast to get shells but did not get any very strong wind.

❶ 這批在美國油鎮採購的機械，足足需3個火車車箱才容納得下，由於到西岸的內陸火車運費太高，乃由東岸紐約裝船，橫過大西洋、地中海，繞蘇伊士運河，運達福州，途中一度失去消息，還以為發生船難；再用砲船轉運至淡水，換上吃水較淺的帆船運到後壠港。

❷ 筆者原解讀機器運到後壠溪出海口南岸的公司寮港；後來發現清末港口仍為出海口北岸的後壠，到日治初期，後壠港淤塞，才遷港至南岸公司寮港（今龍港）。

3月16日

終於讓船灣靠碼頭，先吊起一具鍋爐，才準備將它滾下船，潮水已退，只好等明天再來。華氏58度。

March 16: Had the junkbrought uo to the warfe (wharf) & raised the boiler ready to roal (roll) out when the tide went out will have to wate (wait) until to morrow (tomorrow) Therom (Therm.) 58°.

註 潮汐漲落、風浪大小對帆船進出港十分重要。4月3日，另艘帆船載來剩下的機器，但只卸下部分，即因風浪太大而折返淡水。

3月17日

星期天。將其中一艘船載的機械全部卸下，搬入倉庫。今天是自離開故鄉〔半年〕以來，第一次有「我在工作」的感覺。華氏58度，海風狂掃。

March 17: Sunday Unloaded one junk & stored the machinery away in the store house This is the first day that I have worked since I left home Theom (Therm.) 58° & the wind blowing like the devil.

註 去年9月4日離故鄉迄本日，共六個月又十三天。

3月18日

完成兩船卸貨事宜。預定明天返家。

March 18: Finished unloading the two junks will start for home to morrow (tomorrow).

3月19日

早上8點搭轎從後壠出發，下午1點返抵家門。看見有位彰化來的漢子在村內走動，不知他來深山野村有何貴幹？華氏60度。

March 19: Left Oulan 8 A M (A.M.) with chairs & got home 1 oclock (o'clock) found a Chineman (Chinaman) here from Changhwa dont (don't) know what his business is Theo (Therm.) 60°.

註 本年4月13日記載，此人是發工資的出納小吏。

3月20日

到油泉視察工寮進度，牆壁已築高約2呎。聽說昨天又有兩人被「印地安人」馘首，加起來，一個月內已有6人（原文7人）掉了腦袋。華氏62度，晚間小雨。

March 20: Went up to the oil springs & the new house the walls are up about two feet there was (were) two men killed by the indians (Indians) yesterday that maked seven within a month Theo (Therm.) 62 light rain this evening.

註 3月12日云「已4人遇害」，故加上本日應為6人。3月29日記載，19日另有1人受傷，29日傷重不治。

3月21日

下了一整天的雨，只好待在住處。我覺得今天是我最難受的一天，糟透了！無事可做，無書報可看，更沒有來信可資慰藉，我們緊閉窗戶，點亮寒夜油燈。

March 21: Rained all day so we had to stay in the house all day I think that this is the worst day that I ever put in nothing to read & nothing to do we shut the windows & lit the lamp Therm 62°.

3月22日

壞天氣，在屋內待到下午3點雨停後，下山看大牛車，沒看到任何工人在修車，那裡的張工頭（Chang the foreman）帶工人上山打「印地安人」去了。官方懸賞每殺一位山區土著，可獲賞50銀元。氣溫華氏68度。

March 22: Bad wether (weather) stayed in the house until three oclock (o'clock) then went down to see the wagon no one at work any where Chang the foreman has gone to fight the indins (Indians) the (there) is a reward of $50 offered for every one that is killed The (Therm.) 60°.

註　由行文方式判斷，張工頭應該就是1月20日提到的那位木匠。

3月23日

因為缺乏施工、造車材料，大家都未上工，我們大部分的時間也留在住處。天氣晴朗、暖和，下山一趟去看大牛車，說眞格的，實在去得有點不好意思。華氏68度。

March 23: Stayed in the house all day for the want of something to do wether (weather) pleasant been down to see the wagon until we are ashamed to go any more Ther (Therm.) 68°.

3月24日

禮拜天。天氣分外可人，卻閒得發慌。晃盪上到石圍牆，再折返住處發呆。收到〔本月〕40美元生活津貼。氣溫華氏72度。

March 24: Sunday Very pleasant but nothing to do went up to the stone wall & stayed in the house the rest of the day Received $40.00 on board Ther (Therm.) 72°.

註 3月7日領到當月生活費50美元；4月9日記載領到100美元；5月12日發下50元；因此這40元應是3月的生活費，顯示老王在3月7日只核發一半。但為何少10元？應是代買日用品扣除額，如4月4日記載。

3月25日

上油泉，只有幾個工人（few, 原文用a few）在築路，更少的人（fewer, 原文a few more）在蓋工寮，進度十分緩慢。傍晚在河裡泡澡。華氏75度。

March 25: Went up to the oil springs a few men working on the road & a few more working on the house very slow Went in bathing this evening Ther (Therm.) 75°.

註 小絡英文文法欠佳，他應是把第一個few（很少）誤用成了a few（一些）；第二處a few more正確用法應是fewer，否則工程不會那麼慢。4月2日亦犯同樣的錯誤。此時正是春耕農忙期，兼差工人大部分要下田；夏季農忙期，甚至全部動用士兵取代工人，如7月24日載。

3月26日

下山看大牛車製作進度。收到一副28張的骨牌遊戲（dominoes）。氣溫華氏72度。

March 26: Went down to see the wagon came back & received a set of dominos (dominoes) Ther (Therm.) 72°.

3月27日

上到油泉，一事無成。下定決心，明天到後壠找些工具、器材回來，並寫一封〔抱怨〕信給唐道台。

March 27: Went up to the oil springs nothing done made our mind up to go to Oulan after some tools to morrow (tomorrow) & send a letter to the Taotai.

3月28日

派一士兵帶著給道台的信到基隆（Keelung），囑他順便代買一些咖哩粉、芥末醬、信封信紙。下山到後壠，花3銀元才買到木匠所需工具、材料，再步行趕回家，累壞了。晚間華氏78度。

March 28: Sent a soldier to Keelung with letter to the Taotai and for some curry mustard letter paper & walked down to Oulan for our carpenter tools could not get chares (charges) without paying three dollars so we walked back very tired to night (tonight) Ther (Therm.) 78°.

註 請參閱本年1月29日、2月8日，唐景星可能人在基隆八斗子，或他的招商局在該地有辦事處。這封信顯然有打小報告的性質，因此促成了一些人事調動，參閱4月17、18日。

3月29日

星期六（按星期五），整天在家未外出。天氣很暖和，華氏80度。聽說10天前被「印地安人」殺傷的人，今天傷重不治。

March 29: Saturday (Friday) Stayed in the house all day very warm Ther (Therm.) 80° There was a man died to day (today) that was shot by the indins (Indians) ten days ago.

註 3月20日正確記載應為「迄今六死一重傷」。他寫成傷者先被「射傷」（shot）；但原住民出草基本上不用火槍、甚至不帶火槍，而是埋伏某地、趁對方不注意再用刀、矛攻擊，得手後以鐵刀割下人頭，因此改寫為「殺傷」。當然，出草生死交關之際，仍可能使用弓箭。

3月30日

聽說有位官員要來本村、再到油泉巡視，等了一整個上午也不見蹤影；下午不再等了，冒雨帶著工具、器材下山組裝大牛車。氣溫華氏66度。

March 30: Stayed in the house in the forenoon waiting for a mandarin that we heard was coming here to go up to the oil spring but he did not come so we went down & put the wagon together Ther 66° rain.

❶ 這位官員顯然是老美不認識的大官，否則他們不會恭候。很可能是台灣煤務兼油礦總辦葉文瀾，參閱本年7月15日。

❷ 本日首次將油泉寫為「oil spring」，未加s。稍後在4月18～20日；5月7、11、15、16、18、25、29、30、31日；6月1、11日；7月31日、8月14日、10月9日，也使用spring；其它皆用springs。

3月31日

〔星期天，〕整天在家乾瞪眼。氣溫華氏66度。

March 31: Stayed in the house all day There 66°.

4月1日

無事可做，閒閒在家冇代誌。華氏70度。

April 1: Monday Nothing to do so stayed in the house There (Therm.) 70°.

4月2日

上油泉，只有小貓兩、三隻（原文用only a few, 正確用法應爲only few）有氣無力的在搭蓋工寮、修築道路。傍晚縱入溪中泡澡。天氣暖和，華氏77度。

April 2: Went up to the oil springs only a few (few) men at work in the house & road Went bathing in the creek this evening very warm There 77°.

4月3日

上午，下山搞那輛大牛車；下午返住處休息。接獲通報云，其它的機械全已運抵後壠。氣溫79度。

April 3: Went down and worked on the wagon in the forenoon stayed in the house the rest of the day Received word that the rest of the machinery was at Oulan Ther. 79°.

4月4日

早上8點出發，中午12點抵後壠。才知那貨船只卸下少部分機器，因風浪太大被迫航離〔折返淡水〕。收到台灣府官方代採購的日用品。華氏65度。

April 4: Started for Oulan 8 A M & arrived here at 12 noon found that a boat had been here with our machinery unloaded part of it & had to go away on account of bad wether (weather) Received our groceries from Taiwanfoo Ther (Therm.) 65°.

註 剩下的機器9日才再運到。

4月5日

〔未記載，顯然挫折重重、心情欠佳〕

4月6日

整天在家。馬偕牧師來訪（George L. MacKay），是我近三個半月以來首次看到的老外，他有個教堂離此約3英里（約8.38華里）。

April 6: Stayed in the house all day Received a visit from Mr. G. L. Macky a missionary who has a chaple (chapel) about three miles from here the first foreign man I have seen in three months and one half.

註 絡克只在4月7日寫對馬偕的姓氏，但k未大寫；其餘用Macky或Mckey。馬偕在本日日記記載，「抵達新港，晚間去後壠與溪洲探望簡時及絡克，待了一些時間。」當晚馬偕顯然住宿溪洲庄。

4月7日

禮拜日。北上到（Went up to）馬偕牧師的〔獅潭底〕教堂，上午參與「漢人」信徒的禮拜儀式，午飯後，參加下午的聖經研習，雖然我們有如鴨子聽雷、不懂半個字。馬偕負責的〔北部加拿大長老〕教會有14座教堂、20位本土牧師，其中6位常與他到處旅行、巡視各地教堂。

April 7: Sunday Went up to Rev (Rev. or Reverend) Mr Mackay (MacKay's) chappel (chapel) attended Chinese service in the forenoon took tiffin & attended meeting in the afternoon could not understand one word he has fourteen chaples (chapels) & 20 native preachers six who travel with him from one chaple (chapel) to an other (another).

❶ 依據用詞、方向、距離判斷，此教堂不是位於後龍新港社的教堂（根據筆者實際踏查，原位於後龍鎮新民里的新港社教堂目前已經不存在，舊址連派出所管區也不知道）；而是溪洲油村略東北方的獅潭底（苗栗獅潭鄉），馬偕在*From Far Formosa*書中以Sai-tham-toe稱之。1873年12月底，他的學生許銳在該地傳教，慘遭山區賽夏族馘首。

❷ 該區信徒似不是「漢人」，或大部分不是漢人，因為馬偕早期在偏遠地區的信徒大部分是平埔族；應是道卡斯平埔族新港社移民山區者。小絡未分辨出。

❸ 馬偕當天以腓力比書第4章第11節講道，並載「兩位美國人出席」（The two Americans present）。*Mackay's Diaries*, p. 241；《馬偕日記》，頁333。

4月8日

整天在家，靜候貨船再運機器到來的佳音。

April 8: Stayed in the house all day expecting a boat with our machinery.

4月9日

清晨，被一陣重捶的敲門聲驚醒，推門一瞧，原來是淡水來的波典先生（Mr. P. Boudains）幹的好事。他搭乘一艘砲船，拖來爲求平衡起見、將機械均分裝載的兩艘帆船。共進過早餐，一道趕下山，他旋即搭砲船回淡水，我們則忙著卸貨。收到兩人美金100元的生活費。〔夜宿後壠〕

April 9: Walked (Woke) up this morning & found Mr (= Mr.) P Boudains from Tamsui pounding at the door he came down on a gun boat that towed down two junks with the balance of our machinery he took breakfast with us & returned to Tamsui on the boat Unloaded the junk Received $100 board money.

❶ 波典可能是招商局雇員，因為有官股色彩的招商局才有辦法動用砲艇，何況總辦是唐景星。Sampson Kuo則解讀為洋商經營的輪船公司代表；如是，為何3月15日首次運來未現身？何況洋公司不可能請得動砲船。

❷ 絡克寫的是美金金額，但實際領到的應是當地通行的等值銀元，否則無處兌換、無法購物。

4月10日

將機械搬入倉庫，發現少了一箱，還好無關緊要。借8〔銀〕元給士兵購買食物（chowchow），換來他們答應以後如果早上未下雨，願意到油泉值勤的承諾。

April 10: Stored everything away in the godown found one package short nothing of any account Lent the soldiers eight dollars to buy chowchow with will go to the oil springs in the morning if it dont (don't) rain.

4月11日

早上8點離後壠，花1〔銀〕元僱請3名苦力，幫忙挑運我們採購的食物、雜品，中午12點到家。家裡〔僕人、廚子間〕沒發生狀況，一切安好。

April 11: Left Oulan 8 A M & arrived hame (home) at 12 found everything all right hired three coolies to bring our stores from Oulan for one dollars Ther (Therm.) 80°.

4月12日

上到油泉工地，共約75名工人，有的在修築鑽油處斜坡（grade），有的在搬運山上砍下的木材。天氣十分暖和，華氏80度。

April 12: Went up to the oil springs about 75 men working on the graid (grade) part of the timber (timbers) brought down very warm very warm Ther (Therm.) 80°.

4月13日

那位負責支付蓋工寮苦力工資、彰化城來的漢子，居然沒帶錢來！他說明天會派一士兵回去張羅。華氏80度。

April 13: Went up to the oil springs Chineman (Chinaman) came from Changhwa to pay the men on the house but there is no money send a soldier for some to morrow (tomorrow) Ther (Therm.) 80°.

4月14日

星期天。上到油泉，斜坡已完成，苦力正開始從斜坡處修築通往工寮的道路。

April 14: Sunday Went up to the oil springs the graid (grade) is finished & the coolies are making a road from it to the house.

4月15日

上油泉工地製作一具旋轉磨石（grindstone）框架，明天找位土水師（mason）設法切割一塊磨石。很暖和，華氏80度。

April 15: Went up to the oil springs made a fraim (frame) for a grindstone will have a mason cut one out tomorrow very warm Ther (Therm.) 80°.

4月16日

上午未出門。下午到油泉，途經溪畔，苦力正在修復上次被水沖垮的步橋。找到一位土水師製作一塊磨石（輾石）。

April 16: Stayed at home in the forenoon in the afternoon went up to the oil springs the coolies were repairing the foot bridge that cross the river have one man making a grindstone Ther (Therm.) 75°.

4月17日

上到油泉工地，發現那位石匠已做好磨石、苦力仍在修橋。〔夏〕道台從龜崙嶺（Kulyny）派另一位軍官（officer）到此，負責管理苦力。

April 17: Went up to the oil springs the man finished the grindstone to day (today) coolies working on bridges The Taotai went (sent) another officer from kulyny to take charge of the coolies.

註 Sampson Kuo解讀為唐景星從基隆派一位文官來管理工人；但唐並無人事任用權，因此似應由有調動人事權的台灣兵備道夏獻綸所為。如該員從基隆調來，應屬水師營，不在老樂轄下。公館附近山區為泰雅大湖群汶水部落出沒地，這位軍官的到來顯然不只管理苦力，也有保護油井的考量。另從士兵大量增加、老美改派士兵到基隆採購，可能也與這位「北官南調」有關。詳本年7月1日、2日，使用的地名為Keeluney。絡克常用Keelung稱基隆；這個Kulyny或Keeluney地方，可能是山防重地龜崙嶺（桃園龜山鄉北方山區）。

4月18日

上到油泉工地，將磨石吊置妥當。剛走馬上任的軍官與我們同行，並一道下山共進午餐。〔唐〕道台派了一位新譯員到任，因此這裡共有翻譯、軍官各兩名；聽說王大官人正從台灣府返後壠途中；駐紮彰化的樂副將，預定明天上午抵達。

April 18: Went up to the oil spring hung the grindstoun (grindstone) the new officer came up & came down with us & took tiffin went (sent) a new interpreter so now we have two 〔interpreters〕 & two officers and Weng (Wong) on his way back from Taiwanfoo the colonel will be here in the morning from Changhwa.

❶ 這是愛抽鴉片、喜遲到早退、扣員工薪水、疑似拿回扣的老王，最後一次在台灣歷史舞台露臉演出，此後未再提到他。顯然他並未回任，這可能與絡克3月28日寄出的抱怨信有關。

❷ 1月17日，唐景星帶一位譯員到油村，因此推斷新的翻譯也是他所派任。這位譯員可能是小陳（Chen, Cheng or Chang）或文金（Boon Kin），以小陳可能性較高，詳本年4月26日、6月9日、7月9日；當然，小陳也可能叫小張或小蔣。

4月19日

村民大部分到油泉上工，我們則在住處等老樂，直到中午12點仍不見他的蹤影。下午到油泉兜了一圈，然後回家。

April 19: Waited until 12 noon for the colonel & the rest of the Chinese to go up to the oil spring went up looked around a little & came home.

4月20日

上到油泉工地，簡時與我在9位士兵協助下，把工地的兩座井架基石（derrick sills）架妥。華氏81度。

April 20: Went up to the oil spring & with the help of nine soldiers Karns and I frameed two derick (derrick) sills Ther (Therm.) 81°.

4月21日

〔星期天，〕整天未出門。首次接到來信，而且一口氣來了14封，另有24份報紙，是〔從3月9日以來〕43天內唯一的郵件。晚上傾盆大雨，華氏78度。

April 21: Stayed in the house all day Received 14 letters & 24 papers the first letters I have recd & the onely (only) mail for 43 days Raining quit (quite) hard to night (tonight) Ther (Therm.) 78°.

註 他們去年9月12日從舊金山出發，10月13日抵上海，前後32天；今年11月16日從香港出發返美，12月12日抵舊金山，航程27天。因此，當時美國西岸至清國沿岸的客輪約需30天左右。加上美國陸地運程及轉運到台灣之時程，約需43天。

4月22日

上到油泉，因爲下大雨，木工都沒在油泉工地露臉。架妥一具木輪軸架（jack post）再回家。華氏72度。

April 22: Went up to the oil springs but the carpenters did not come on account of the rain framed one jack post & came home Ther (Therm.) 72°.

4月23日

上到油泉工地，有3位木匠在工作；樂副將從後壠來到此地。華氏78度。

April 23: Went up to the oil springs three carpenters came to worke (work) to day (today) the colonel came back from Oulan Ther (Therm.) 78°.

4月24日

中午樂協台的朋友宴請我們，酒足飯飽的他就在我們的床上呼呼大睡，酣聲足足響了整個下午。

April 24: The colonel took dinner with us after eating a Chinese dinner at a Chinese friend of his then he went to sleep on our bed (beds) & snored all afternoon Ther (Therm.) 80°.

註 由此處看得出兩位老美合住在小廟的一間側房；而老樂塊頭很大，兩張大床才容納得下他，亦或他醉得把兩床合併、倒頭橫睡？

4月25日

一整天都在油泉工地搭設井架基石。老樂送僕人阿三一套白色衣服。某人邀宴，我們予以婉謝。寫信給家人Flora, Dave, Irvine, Wheeler。

April 25: Went up to the oil springs stayed all day working on sill the colonel presented Ah Sun with a white suit Received an invitation to a Chinese dinner but declined wrote letter (letters) to Flora Dave Irvine & Wheeler.

註 他們與文官、村民的互動似乎欠佳；但和武官、士兵相處頗佳。如3月4日士官長夫人來訪、13日與士官長看戲、4月10日借錢給士兵、18日替新到任軍官接風，與老樂的友善關係更不在話下；但卻不能與老百姓常相往來，註定了在台寂寞、苦悶、無法排解的日子。

4月26日

到油泉工地架設鑽機（rig），將全部泥製基石（mud sills）安置好，主基石（main sill）也已備妥、隨時可安放。接到後小陳（Chen）來信，催我們趕緊把機器運上山。華氏80度。

April 26: Went up to the oil springs worked on the rig placed all the mud sills & have the main sill ready to place received a letter from Chen to go to Oulan and move the machinery up Theom (Therm.) 80 °.

註 會寫英文信的小陳應是絡克寫信抱怨後，唐景星再派的新上任翻譯，幹勁十足，還會催他們上把勁；相對於3月14日、不肯走遠路下山的油條譯員（可能是6月9日的文金），兩者工作態度相差不可以道里計。

▲抽油桿式抽油機（陳政三攝）

▲馬偕與他的小博物館研究室（後龍「城外基督長老教會」陳列的馬偕像，陳政三翻拍）

▲後龍溪口龍港（陳政三攝）

▲後龍基督長老教會（後龍南龍里城外98-1號，不是已廢的新港教會原址，陳政三攝）

三部曲

幾滴石油．幾滴血淚

第一滴油

絡克有次在今明德水庫附近的茄苳坑（Eltarcau, 頭屋鄉明德村），遇到5位下山交易的賽夏人對他說，小絡很像他們祖先其中的一支，使得小絡覺得自己好像是個「老罪人」（old sinner），打趣的寫道有機會的話，要翻查土著的族譜（pedigree），假如發現所言是事實，立即上吊自殺。據絡克晚年回憶，獵頭族（head-hunters）只殺漢人或敵對部落人，對白人則無敵意。有次受邀到一個部落，土著聽到絡克裝煙草說出「tobacco」時，驚訝地表示與他們的用詞相同，酋長立刻宣稱小絡是他的gip kik——好兄弟、好朋友。但小絡在日記中未曾提到訪問過部落，而且不論賽夏族或泰雅族稱呼朋友、兄弟皆非使用該字，他可能吹牛，而且胡扯出那個字。

官方爲了防範土著出草，每月付50銀元「保護費」給附近部落，要後者擔任保全工作，監視、降低其它部落的覬覦。可是駐紮油井工地的士兵曾要求老美分出3名貼身保鑣上去保護他們，以免遭到「印地安人」攻擊。美國佬聞言，啼笑皆非，回話建議那百名米蟲「勇士」（brave men），爲了保住小命，最好移營到安全之地。這可得罪了軍爺，所以有次老美像傻瓜似的冒雨上工，傍晚下工時河水暴漲，步橋又早已毀壞，無法渡河；如果折返油泉工地，那裡的士兵恐怕不會歡迎他們，只能另謀下榻之地；他們先向村民買隻雞、幾個甘薯充飢，再爬過一座1,000呎的山，終於找到某漢人的茅屋借住一宿，次晨河水水位稍退，這才游泳過河、濕淋淋地回到溫暖的家。

1878年4月9日，所有機器都已運抵後壠，除了那輛三條牛拖拉、運載蒸氣鍋爐的大牛車（big wagon），另外僱了14輛兩輪牛車（Chinese carts），共運載44輛次，才把3個火車車廂容量的機器全數運送上油井。5月上、中旬，唐景星的特別助理布郎（R. M. Brown）來到油井附近幫忙，與簡時、絡克同甘共苦，協助搭建井架、安裝機器。絡克不但披頭散髮，而且奇裝異服，5月12、17兩日記載穿著條紋睡衣（pajamas）上工，顯然已快無衣褲可穿，所以拜託要回基隆的布郎代買一套衣服、幾雙鞋子。這足以媲美1874年隨日軍來台採訪的《紐約前鋒報》（New York Herald）記者豪士（Edward House），後者身穿條狀睡衣褲、頭頂草帽，還手撐陽傘，腳蹬雙草鞋，這樣的裝扮是他一向自認最適合記者在熱帶地區採訪的工作服，西鄉從道將軍居然很羨慕，語帶幽默的恭喜豪士有機會穿睡衣上戰場。

5月19日，老美買了一些布料製作美國國旗，先將布剪好，再交給士兵眷屬按照樣式裁製。30日，國旗做妥，小絡冒雨要士兵揹他渡過高漲的河水，不知有意或無意，他寫道，「士兵再次滑倒，又讓我摔進河裡。但還是想辦法過河，美國國旗終於高掛井架上、迎風飄揚」。這面國旗不是當時台灣唯一的一面，因為同年3月20日，兼任美國駐淡水副領事的茶商德約翰（John Dodd），比他們先在台灣升起了美國國旗。事實上，德約翰也是第一位試圖開發台灣石油的洋人。根據《淡水廳志》、《苗栗縣志》載，同治四年（1865）德約翰每年以千餘圓，向客籍通事邱苟租得貓溪頭內山（出磺坑附近）的採油權，引起同治三年以每年百餘圓承包的原吳姓租戶不滿，連年互控、集眾械鬥。1870年3月淡水廳逮捕邱苟、殺之，中斷了德約翰成為「台灣石油大王」的美夢；但卻讓他的大名標在當時英、法軍事地圖上，雪山主峰西南邊的山脈（可能是小雪山、中雪山、大雪山連成的山脈）被命名為「德約翰山脈」（Dodd's

Range）。

8月3日終於鑽打到油層，但流出的鹽水卻遠多於石油。絡克特別在該日日記標上年代，以紀念全清國，也是台灣地區用機器鑽出「第一滴油」的大日子。不過該年基隆煤炭生產量大減、品質欠佳，導致缺煤的困擾仍然存在，使鑽油工程時斷時續，他們可能也使用到木材、木炭當燃料。8月22日，樂文祥居然故意縱火，把他們在溪洲庄的廚房燒毀，迫使老美搬到油井工寮。23日小絡記道，「今晚，終於有了自從抵達大清帝國後，擁有第一個稱得上屬於自己的『家』。好個喬遷之『喜』！」

無聊的日子難熬，他們又無法或不願意融入在地人的生活圈。日記中，過年過節未見他們受邀到村民家中吃拜拜，顯見彼此互動不多、關係不是挺好的。4月25日，有人邀宴，也遭他們婉謝。6月5日、農曆五月初五端午節，日記居然未提到粽子，美國「阿逗啊」不吃台灣豬肉，至少也可叫廚子、僕人改用雞肉當餡嘛。日記也未曾對村民或工人有較深刻的描寫，在在顯示他們與當地人士的隔閡與疏離感。

單身的小絡偶爾釣魚、游泳、看賣膏藥的雜耍團表演以茲消遣；簡時「幸有意中人堪尋訪」，但訪出了亂子，問題不在他已婚身分，而是他腳踏兩條船，劈腿劈到廚師的兩位女兒，真不知「兔子不吃窩邊草」的道理。小絡只是輕描淡寫，稱「簡時與廚師的女兒們之間發生了點麻煩事（trouble）」，未提到最後如何解決，想必仍是「始亂終棄」老戲碼重演。

9月3日，他們完成了與清國官方簽署的一年合約。但鑽井的工作十分不順利；馬偕牧師再度來訪受阻於湍急的溪水，而不相干的漢人卻相繼湧到油井「觀光」；有事沒事還得與漢人「異教徒」鬥嘴、賭氣，根據合約老美對工人不得有凌辱或處罰的行為，因此被煩得要死；嚴重缺乏人手之際，米蟲士兵卻整天閒晃，看得小絡火冒三丈，卻無計可施；麵粉又常缺貨，連家書也至少一個多月才收得到，精神、物質皆極度貧乏。諸事不順情況下，風土病悄然來襲。7月22日，兩位士兵死於瘧疾、飲水不衛生所引起的傷寒、森林熱。據1878年〈淡水海關年報〉，當年風土病橫掃北台灣，洋人染上有嘔吐現象，漢人則無，不加治療足以致命。稍後疫情蔓延，9月14日油井附近的人絕大部分染病，15日簡時，22日絡克相繼生病。但合約規定洋技師生病須自行就醫，意即「萬一病死，官方不負責任」，這更加深他們不再續約的念頭。

9月27日，病懨懨的絡克搭轎永別出磺坑油井，29日抵達大稻埕，心情爲之大好。簡時多待在油井10天，稍後也是抱病北上。就此結束此段台美不愉快的合作經驗。根據當年〈淡水海關年報〉，這次鑽井共抽出約400擔（8,800加侖）石油，其中100擔供糖廍照明用，其餘300擔儲存於後壠。如以絡克草估清國投資了10萬美金計算，那麼每加侖石油（4.55公升）成本高達11.36美元。

1878年（光緒四年）

4月27日

下山到茄荎坑（Eltarcau, 頭屋鄉明德村，明德水庫附近），遇到5位從山區下來呼吸文明氣息的「印地安人」（Indians），說我很像他們祖先其中的一支，聞言頓覺自己好像是個前科累累的「老罪人」（old sinner），但我沒說什麼、也無話可說，接過他們遞過來的酒就喝，下定決心有機會的話，翻查他們的族譜（pedigree），假如發現所言是事實，立即上吊自殺。

April 27: Went down to Eltarcau and see five indins (Indians) who had come down to get civilized they informed me that I was one of ther (their) ancestors it made me feel like an old sinner but I didnt (didn't) say anything I did not have anything to say I drank the wine that they gave me & resolved that I would look up ther pedagere (pedigree) & if I found out that it was true I would go & hang myself.

❶ 茄荎坑在出礦坑與溪洲油村北方、後東偏南方，後龍溪支流老田寮溪北。他未交代為何到該地，可能去看油桶存放處。茄荎坑北方不遠處即是錦水，後來也發現蘊藏豐富的石油、瓦斯。參閱本年6月3日、7月28日。附近的三灣、南庄、獅潭三鄉昔日皆為賽夏族居住地，他遇到的可能是賽夏族。

❷ 據絡克晚年回憶，獵頭族（head-hunters）只殺漢人或敵對部落人，對白

人無敵意。有次受邀到一個部落，土著聽到絡克裝煙草說出「tobacco」時，驚訝地表示與他們用詞相同，酋長立刻宣稱小絡是他的gip kik——好兄弟、好朋友。但日記未曾提到訪問過部落，而且不論賽夏、泰雅族稱呼朋友、兄弟皆非用該字，不知他可有加油添醋？

4月28日

獲報道路已大致修通，還說最好趁雨季之前，儘快把機器搬運到山上，以免路基遭雨水沖失。一早立即下山，想趕到後壠搬運機器，才到半途，發現沿途道路幾乎尚未動工、顛簸難行。眞不知該如何是好，仍到後壠去呢，或折回家？他們爲甚麼要騙我們？想了想，最後還是下山〔到後壠〕。

April 28: Sunday Left home this morning for Oulan to move up the machinery as the Chinese informed me that the road was all right & that they wanted to move the machinery before the rain would wash the roads out came down here & found that thay (they) had done nothing to the road dont (don't) know what I shall do now wait or go back how thay can lie walked down.

4月29日

整天意興闌珊，〔待在後壠〕不想動。魏先生（Mr. Hugh）來找我，稱他已找了12名工人修路。半信半疑的出門前去瞭解，好個老魏，他眞辦到了。

April 29: To day (Today) did not do anything Mr. Hugh came to see me & said that he had set 12 coolies at work repairing the road I went out to see & I was surprised to find that he had 〔done that〕.

❶ 此時可能農忙，工人難找。

❷ Sampson Kuo於論文（頁158）解讀Hugh是官方聘請、協助鑽井的「洋雇員」，但各種資料均未顯示這一點。由5月29日記載他的漢語流利，顯見他是漢人；再由他找修路工人、6月3日支付絡克生活費，極可能是取代老王職務的胥吏（未經正式考試的小吏）。

4月30日

今晨，僱了10輛兩輪牛車（Chinese carts），裝上機器要他們運到油泉工地；他們掂斤估兩、專挑輕的載，對重的避之唯恐不及。但他們沒得挑，必須聽我的。天氣十分悶熱，跳蚤相當多。

April 30: This morning hired ten Chinese carts & loaded them and sent them to the oil springs thay weighed everything that thay (they) could lift the heavy things thay had to take my word for Very hot lots of fleas.

註 Sampson Kuo將Chinese carts解讀為人力車。是否？詳5月12日。

5月1日

上午待在後壠住處。下午，7名苦力驅趕由三條牛拖拉的大牛車（big wagon）下山來到後壠，準備運載蒸氣鍋爐。寫封信通知〔人在油村的〕簡時。

May 1: Wednesday Stayed in the house in the fore noon (forenoon) this afternoon the Chinese brought our big wagon down Seven coolies & three buffo (buffalo) to bring it wrote a letter to Karns.

❶ 這輛絡克設計、剛完成的美式大牛車，顯然有別於傳統一頭牛拉的台式兩輪牛車。他對大牛車外型無任何著墨，可能類似美國西部開拓史用的加大型無篷四輪馬車。

❷ 絡克寫信通知簡時，透露後者留在山上照料。

5月2日

修理大牛車，換掉一些車輪、螺栓，頗費時。北風強勁、陽光普照。

May 2: Repaired the wagon put some bolts in & fixed the wheels & quit (quite)? strong north wind with plenty of sun.

5月3日

今天下著毛毛雨，苦力仍上工修路。晚間，布郎（R. M. Brown）抵達後壠。

May 3: coolies at work on the road to day (today) Light rain R M Brown arrived here this evening.

註 布郎先後在後壠、油村停留半個月，5月17日返基隆。他在本年中、迄年底似乎一直待在基隆招商局辦事處。

5月4日

與布郎指導7名苦力修整道路。

May 4: Mr. Brown & I supertended 7 coolies that was (were) working on the road.

5月5日

〔星期天，〕在布郎及苦力協助下，把較小的機器裝上14輛小牛車，運載到油泉。

May 5: With the help of Mr. Brown & coolies we loaded 14 carts & sent them to the oil springs.

5月6日

把笨重的一具鍋爐裝上大牛車，上午11點啓程運往油泉（oil

springs），行駛狀況看起來十分良好。我與布郎另行僱轎，下午4點回到油村，立即再趕路上油井（oil well）。

May 6: Loaded the boiler on the wagon and started it for the oil springs 11 oclock (o'clock) it runs very well procured chairs and with Mr. Brown came back to the oil springs arrived here at 4 P M (P.M.) went up to the oil well.

註 機器逐漸運上山，工事有了眉目，絡克因此首次使用「油井」（oil well）。6月4日第二次使用油井字樣。

5月7日

在油泉工地開始搭建井架（derrick），我與布郎把井架搭高4呎，簡時組裝機器。傍晚返家途中，3人在河中深水處洗澡。

May 7: Went up to the oil spring and commenced to build the derrick run it up 4 feet Mr Brown & I have (had) done the work in the derrick & Karns furnished the material Took a bath in the deep hole on the way home.

註 Mr.的寫法，絡克常忘了標上句號；以下未標者，為原字呈現，不再註明。

5月8日

到油泉完成井架，布郎在塔下幫忙。天氣十分暖和，華氏80度。

May 8: Went up to the oil springs finished the derrick Mr Brown worked on the ground very warm 80°.

5月9日

裝設鑽機（rig），布郎和我合做一張吊椅（chair）。華氏82度。

May 9: Worked on rig Mr Brown & I built a chair Thom (Therm.) 82°.

5月10日

老魏上油泉來找我，稱缺材料、器具，工事進展不順。華氏82度，今天與布郎在河裡洗了3次澡。

May 10: Went up to the oil springs Mr. Hugh came up did not do much did not have anything to do with Theom (Therm.) 82 took three baths in the river with Mr Brown.

註 老魏的窘狀，參閱5月15日日記。

5月11日

到油泉做好吊椅，著手開挖導管洞穴（conductor hole）。用繩子在井架四周圈圍，隔離四周圍觀的村民。天氣燠熱，室內溫度88度。

May 11: Went up to the oil spring finished the chair commenced a conductor Hole put a rap (rope) around the rig to keep the Chinese out of the way Very warm 88 °in the shaid (shade).

5月12日

星期日。穿著條紋睡衣（pajamas）上到油泉，但沒動手做事。從後壠來的4輛二輪牛車抵達，加起來共已載44輛次。華氏90度（攝氏32.2度）。收到生活費50美元。

May 12: Sunday Went up to the oil springs in my pagaimas (pajamas) did not not do anything Four carts came up from Oulan which makes 44 in all Theom (Therm.) 90 ° Recd $50.00 on board.

❶ 本日與7月5日、6日、8日、16日，均為華氏90度，是在台期間有紀錄的次高溫。6月12日室外溫度居然高達華氏104度（攝氏40度），有可能為筆誤。

❷ 如用人力拉車，44輛次無法載完需3個火車廂容量的機器，顯然是用牛車。

5月13日

在油泉操作井架安置引擎。下午天公落水；晚間減弱成毛毛雨。

May 13: Went up to the oil springs worked on rig set the engine Rain in the afternoon & light rain this enening (evening).

5月14日

把蒸氣鍋爐吊置妥當。晚上與村民同賞來村賣膏藥的雜耍團（circus）。

May 14: Worked on rig set the boiler Had quit (visit ?) a circus with the Chinese in the evening.

5月15日

老魏上來油泉工地找我們，解釋因為無錢購買材料，所以沒辦法照我們的意思把工作做好。

May 15: Went up to the oil spring Mr Hugh came up to day (today) said the reason that he would not furnish what we wanted was that he had no money.

5月16日

白天在油泉吊置一些機器。晚間我與布郎繞小鎮（town）散步，在某處漢人舉行廟會（joss pageant）的地方佇足觀賞，後來加入祭拜行列，可能我們舉止太活潑，又拜得不三不四，他們似乎很高興看到我們離開。

May 16: Went up to the oil spring worked on rig In the enening Mr Brown & I took a walk around the town stoped (stopped) to a place where the Chinese were having a joss pigeon (pageant) we stoped & took part but we made it so lively for them that thay (they) were glad to get rid of us.

❶ 之前，絡克不曾使用「小鎮」形容油村，但他們顯然並未離村到其它地方。

❷ 簡時38歲、布郎32、小絡28，後兩者年齡較接近，因此較常在一起行動。

5月17日

布郎今晨回基隆，遞過20美金，託他爲我買一套衣服、幾雙鞋。仍穿著睡衣上油泉，督導工人挖掘導管洞穴。晚上雷雨交加，華氏85度。

May 17: Mr. Brown went back to Keelung this morning I gave him $20.00 to buy me a suit of clothed & some shoes Went up to the oil springs in my pajamas and looked ay coolies while thay (they) dug in the conductor hole Thunder storm this evening 85°.

5月18日

渡河前往油泉時，〔後龍溪〕溪水暴漲而且混濁，士兵揹我過河、滑了一跤，把我拋入河裡，只好濕淋淋地折返。雨勢更大。

May 18: Started for the oil spring the river quit (quite) high & dirty the soldier that was carrying me sliped (slipped) & throwed (threw) me in so I came back more rain.

5月19日

星期天。買了一些布料做美國國旗，將布剪好，交給士兵眷屬按照樣式裁製，心情十分愉快，氣候也很晴朗、舒適，華氏80度。

May 19: Sunday Bought some cloth to make an United States flag cut it out & set the soldier (soldiers') wives at work on it quit (quite) pleasant to day (today) 80°.

5月20日

在油泉工地架妥煙囪。導管洞穴積滿水，先將水汲出、再挖深1呎，聞到些微瓦斯氣，還冒出許多石油。潮濕的天氣使得洞穴嚴重塌陷（cave in），我們決定明天早上將導管（conductor）插入洞穴、儘可能鑽入地底。

May 20: Went up to the oil springs put the smoke stack up the conductor hole was full of water bailed it out & dug it about a foot deeper struck some gas and considerable oil the wet wether (weather) makes it cavein (cave in) so bad that we have concluded to put in the conductor in the morning & drive it the rest of the way 78 °.

5月21日

挖深洞穴到不能再挖爲止，足足有19呎深，地下水不斷湧出，將汲水管（water pipe）插入，再鋪妥鑽井鐵架塔台（derrick floor）。陰天、微雨，華氏78度。

May 21: Dug the conductor hole as deep as we could - 19 feet - on account of water put the pipe in & laid the derrick floor finished laying the water pipe wether (weather) cloudy light rain 78 °.

5月22日

早上醒來，推窗一看，外頭下著豪大雨，河水暴漲，是上山迄今以來最大的雨勢，一直下到晚上8點才停止。

May 22: Walked (Woke) up this morning and found it raining hard and it has kept it up all day it is the heaviest rain fall (rainfall) that we have had since we have been here the river is very high To night (Tonight) 8 P M it has stoped (stopped) raining & is clearing off.

5月23日

大雨續下，只好再留住處一整天。晚上雨停，但仍烏雲密佈，天氣很熱、華氏82度。100名駐紮油泉的士兵（government soldiers）要求派3名我們的兵（our soldiers）上去保護他們，以免遭「印地安人」攻擊。我們回話表示，為了保住小命，他們最好移營到安全之地。開什麼玩笑，要我們分出僅有的11名兵中的3個，去保護百名米蟲「勇士們」（brave men）！

May 23: More rain so I had another day in the house to night (tonight) it has stoped (stopped) raining but it is very cloudy & very warm 82 The 100 government soldiers that is (were) located at the oil springs ask (asked) for three of our soldiers to come & guard them as thay (they) are (were) afraid of the indins (Indians) we sent them word that thay had better remove to a place of safety as we would hate to have any of them hurt 100 brave men & that we could not spare any of our 77 (11) soldiers to guard.

❶ 絡克晚年回憶，他們原有官方派的20名保鑣（bodyguard or personal soldiers），後剩11位，部分攜眷。這些保鑣可能是樂文祥所派的、陪同他們的士兵。1月8日，老樂派9名親兵陪他們到油村，往後陪上山、到工地、赴後壠的兵都是住在溪洲油村的士兵，而非駐紮出磺坑油泉的士兵。樂文祥於4月23日再到油村，可能加派11兵，總共20位士兵，但後來小村住不下攜家帶眷的人數，乃減成11名。

❷ 絡克留有一份11位head soldiers的名單，或許因此Sampson Kuo認為他們可能就是小絡稱的保鑣。不無可能；但筆者仍認為head

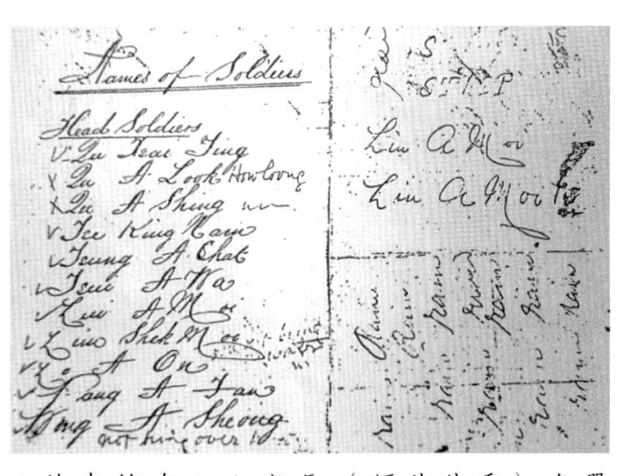
Names of Soldiers

Head Soldiers
Lu Tsai Ting
Lu A Look How long
Lu A Shing
Tce King Nam
Tsung A Chat
Tsui A Wa
Liu A Moi
Liuo Shek Moi
A On
Fang A Fan
Wong A Sheong
not living over 10

Liu A Moi
Liu A Moi

▲絡克持有之士官長（額外外委）名單（Sampson Kuo書，陳政三翻拍）

soldiers是帶兵的「額外外委」（士官長），參閱3月4日、13日。

❸ 他們的回話得罪了「勇士們」，導致7月31日的慘況。

5月24日

昨夜整晚下著生平僅見的傾盆大雨，早上雨停，幾位附近居民在下游撈起斷成數節的抽油桿（sucker rods），送還給我們。白天未再下雨，河水從早上暴漲，到晚間已驟降10英尺。

May 24: Rained the hardest I ever heard all through the night some of the natives brought us peaces (pieces) of sucker rods that they had found floating down the river There has been no rain during the day & the river has fell (fallen) ten feet since morning.

5月25日

今天雨停，高漲的河水仍未退，3名士兵冒險渡河、上油泉，回報停放工地的大牛車已被大水沖走，連帶放在車上的32節抽油桿也不見了。

May 25: No rain to day (today) but the river was to (too) high to be crossed three soldiers went up to the oil spring thay (they) report the big wagon gone & 32 sucker rods with it Thereon (Thermometer) 80°.

註 這3位士兵即是絡克晚年口中的保鑣，也是陪他們上山下海、借錢買菜的阿兵哥。

5月26日

星期天。想到石圍牆庄（the Stone Wall）一趟，河水太高，只好作罷。收到Flora, Bovaried, Seyfang的來信。寫信給Alase（可能是Alice之誤）與Mr. H. W. Locke。

May 26: Sunday Went up to the stoan (stone) wall but the water was to (too) high 〔to〕 cross Received a letter from Flora and one from Bovaried one from Seyfang wrote a letter to Alase (Alice) & one one to H. W. Locke.

❶ 小絡父親的名字叫E. G. Locke, 一位伯叔Jonathan, 日記所提之人應是長輩，所以冠上Mr., 可能是另一伯叔。

❷ 美國南北戰爭時期，南軍猛將General Thomas Jonathan Jackson（1824.1～1863.5）綽號「石牆將軍」（General Stonewall; Stonewall Jackson）；可能因此，絡克把石圍牆庄譯成他的別名。

5月27日

上到距離油泉半路的溪邊，河水太高，不得不再折返。寫信給Bovaried, 收到布郎來信，稱唐道台即將到此地住一陣子。氣溫華氏90度。

May 27: Went part way to the oil springs but the water was so high that had to come back wrote letter to Bovaried Thereometer (Thermometer) 90° Received letter from Brown stating that the Taotai was coming down here to stay.

5月28日

整天飄著毛毛雨，大部分時間待在屋內。今天有拜拜，村民忙著祈神燒香。華氏85度。

May 28: Light rain to day (today) stayed in the house most of the day There was a ? to day and the inhabitants have been having a holly (holy) day. There (Therm.) 85°.

註 本日為農曆四月二十六日，五府王爺（千歲爺）生。

5月29日

想上油泉，但河水過高，只好折返。老魏來看我，他和小陳用漢語

（Chinese）談了兩個半小時。在房內殺死一隻蜈蚣。

May 29: Started to go up to the oil spring but the water was so high that I had to come back Mr. Hugh came to see me stayed 2and1/2 hours talking Chinese to Chen killed a centeepead (centipede) in my room.

註 這證明Mr. Hugh是個漢人。只是不知講的是閩南、客家、或廣東話？應非北京官話。

5月30日

前往油泉，士兵揹我過河，再次滑倒，又讓我摔進河裡。但還是想辦法過河，美國國旗終於高掛井架上、迎風飄揚。

May 30: Went up to the oil spring in carring (carrying) me acrosst (across) the creek the soldiers fell & throw (threw) me in But the U.S. flag in the derrick.

❶ 同期間至少有兩面美國國旗，每天在台灣的天空迎風搖曳。據〈1878年淡水海關年報〉，該年3月20日，美國駐淡水副領事德約翰（John Dodd）在淡水升起美國國旗。英國籍的蘇格蘭人、「台灣烏龍茶之父」德約翰，文獻上使用過德約翰、力絨士、突得來、突德來、突來德、委員突、突等名字，筆者之前使用「陶德」稱之。他從1868年6月～74年5月兼任美駐淡水副領事，1877年夏再度兼任。他還是開發台灣石油的先趨，《淡水廳志》、《苗栗縣志》載同治四年（1865）他每年以千餘圓，向客籍通事邱苟租得貓

▲德約翰（John Dodd）（原載《歷史月刊》201期）

溪頭內山（出磺坑附近）的採油權，引起同治三年以每年百餘圓承包的原吳姓租戶不滿，連年互控、集眾械鬥。1870年3月淡水廳逮捕邱苟，以邱曾勾引「生番」殺人老案，就地正法，中斷了德約翰成為「台灣石油大王」的美夢；但卻讓他的大名標在當時英、法軍事地圖上，雪山主峰西南邊的山脈（可能是小雪山、中雪山、大雪山連成的山脈）被命名為「德約翰山脈」（Dodd's Range）。先後曾任美國領事的李仙得（Charles Le Gendre）、達飛聲（James Davidson）則將德約翰「發現」石油的年代分別記為1868與66年；不過根據德約翰在一篇刊於1885年《皇家亞洲學會海峽分會會刊》（*Journal of the Straits Branch of the Royal Asiatic Society*）, 15（1885）: pp. 69-78的文章 "A Glimpse of the Manners and Customs of the Hill Tribes of Formosa,"（p. 69）稱他在1865年春季，於雪山及德約翰山脈附近「發現」了油井（the "Petroleum Wells" discovered by myself）。絡克、簡時在返美前，曾住大稻埕、淡水，有機會向老前輩德約翰請益。詳本年10月初日記。德約翰（陶德）1838年生於英格蘭西北部的Westmoreland（後來納入county of Lancashire）, 1864年至淡水定居。參閱〈1864年英駐台灣副領事貿易報告〉, p. 276, IUP, China 6, p. 480; Dodd, *Journal of A Blockaded Resident in North Formosa, During the Franco-Chinese War*, 1884-5, p. 173; Dodd, "Formosa," *The Scottish Geographical Magazine*, Vol. 11, 1895, pp. 553-570；陳政三譯著，《泡茶走西仔反》，頁135註39。1867年顛地行破產，同年5月轉為怡和行代理商。原皆稱其為蘇格蘭人，根據Niki Alsford最新的考據，德約翰於1838年10月25日出生於英格蘭西北的Preston Lancashire；1893年8月與威爾斯女子Mary Lloyd結婚，遂定居北威爾斯，2人無子嗣，1907年7月15日卒於北威爾斯康威縣自治區（Conwy county borough）Trefriw小村。他在台灣曾與女子Taihee（代喜？）生下2子Valentine與Elaine。參閱氏著*The Witnessed Account of British Resident John Dodd at Tamsui*, pp. 5, 29, 30。究竟他係蘇格蘭裔或英格蘭人，待考。有可能是蘇格蘭裔的英格蘭人，就像台灣多地的「閩南化的客家人」一樣。另據《風中之葉》作者藍柏（Lambert）告訴筆者，Dodd祖先原居地在約200年前屬於蘇格蘭，而Taihee有可能是香港當地人。

❷ 《淡水廳志》（頁338）、《苗栗縣志》（頁108）上載的貓溪頭內山，即是《苗栗縣志》內載的磺油山（頁23）、牛鬪山（頁110）；也即《清季臺灣洋務史料》的牛頭山（頁5、21）、牛琢山（頁28）。外國探險家又稱這個地方為Kow Kow, 筆者推測公館附近昔時稻田有許多青蛙，故俗稱「蛤仔市」，即青蛙叫聲——Kow Kow衍生出的地名。

5月31日

老魏今天又上山來找我們，說了很多話、滿腹牢騷，但光是抱怨也無濟於事。到油泉蓋一間收納大皮帶輪的小屋（band wheel house），一度下起大雨，幸好未持續很久。華氏83度。

May 31: Mr Hugh came up to day (today) done considerable talking but it did not amount to anything Went up to the oil spring built band wheel house Rained quit (quite) hard but did not last long There (Therm.) 83°.

6月1日

星期六。昨晚下大雨，河水又暴漲，無法渡河前去油井。晚上天氣涼爽，氣溫76度。

June 1: Saturday Hard rain last night could not get up to the oil spring pleasant to night (tonight) Ther (Therm.) 76°.

6月2日

〔星期天，〕上午去釣魚，下午返家休息。華氏78度。

June 2: Went fishing in the fore noon (forenoon) stayed in the house the rest of the day There (Therm.) 78°.

6月3日

到油泉上工，沒完成多少事。老魏從茄苳坑（Eltarcau）來，付我50美元生活費。氣溫華氏80度。

June 3: Went up to the oil springs did not do much Mr Hugh came up from Eltarcau payed me $50.00 in board There (Therm.) 80°.

註 茄苳坑可能是油桶存放處，參閱4月27日、7月28日。老魏顯然是取代王大官人職務的小官吏。

6月4日

上油井發動蒸氣引擎，運轉良好。氣溫82度。

June 4: Went up to the well got up steam & run the engine works well Ther (Therm.) 82°.

註 自本年5月6日以來，第二次使用油井（well）字樣；以後則經常出現。

6月5日

星期三。今天〔農曆五月初五〕是漢人的全國性節日〔端午節〕，大家都休息，所以我也整天待在家裡。寫信給S. Newkirk及S. B. Logan。華氏82度。

June 5: Wednesday To day (Today) is national holyday (holyday or holiday) with the Chinese no one at work so I stared at home all day wrote to S. Newkirk & S. B. Logan Ther (Therm.) 82°.

註 小絡居然未提到粽子，老美不敢吃台灣豬肉，至少也可叫廚子、僕人改用雞肉當餡嘛；而與他們互動極少的村民，顯然也未送粽子給他們。

6月6日

上油井發動引擎，開始鑽油井。派一兵帶幾封信件到後壠。毛毛雨，華氏80度。

June 6: Went up to the well fired up & commenced spuding (spudding) sent soldier to Oulan with letters Light rain There (Therm.) 80°.

6月7日

到油井上工，但缺煤發動引擎，能做的不多。返程途中遇見4位「印地安女人」，她們下山找漢人〔以物易物〕，正要回家。雨天，華氏80度。

June 7: Went up to the well had no coal so we did not do much coming home met four indin (Indian) women who had been down to see the Chinese & were going back Rain 80°.

註 〈1878年淡水海關年報〉（pp. 216-217或 p. 總345-346）記載，該年基隆煤炭因缺少蒸氣動力挖煤及儲煤處，加上沒有便利的運輸方式，造成生產量大減、價格昂貴。因此，絡克在1月12日、2月26日提到他四處找煤的事；1月14日認為瓦斯替代能源應該夠用，但顯然事與願違。

6月8日

星期六。整天雨勢不斷，只好待在廟中屋內。據說老魏已經去了（has gone to）基隆。華氏80度。

June 8: Saturday Rained all day stayed in the house Reported that Mr Hugh has gone to Keelung 80°.

註 這是老魏最後一次在日記出現。他用has gone to述說老魏「去」基隆；但在描述從基隆來的唐景星（1月29日）、布郎（5月17日）「回」基隆時用went back to, 似乎透露老魏不是從基隆派來的。

6月9日

〔星期天，〕在家休息，寫信給Charles Beown（可能是Brown之筆誤）。小陳（Cheng,可能是Chen筆誤，反之亦然）從後壠上來，文金（Boon Kin）則到後壠去。

June 9: Stayed at home Wrote a letter to Charles Beown Cheng came from Oulan & Boon kin went to Oulan.

註 兩人交班輪流陪老美，看得出應是翻譯。參閱4月18日。

6月10日

整天在家，寫信給Nett及Dave。到基隆去的幾名士兵返回，帶回一雙鞋、煙草、雜貨等物。華氏82度。

June 10: Stayed at home all day wrote a letter to Nett & one to Dave Soldiers came back from Keelung brought a pair of shoes smoakinning (smoking) tobaca (tobacco) groceries & Ther (Therm.) 82°.

註 鞋子是5月17日託布朗代買。

6月11日

全天大部分的時間都下雨，到油泉上工。聽說馬偕來找過我們，但溪水太高，無法上到油泉工地。

June 11: Went up to the oil springs Rained most of the day Mickay (MacKay) Called to see us but the water was so high that he could not get to the oil spring.

註 馬偕甫於5月27日與五股坑的張聰明（蔥仔）結婚；他常與太太四處旅行傳道，根據新版的《馬偕日記》（頁339-340）；*Mackay's Diaries*（p. 245），他攜新婚妻子蔥仔來了趟「石油蜜月之旅」，另有三位弟子隨行，因大雨，無法進入山區。

6月12日

星期三。到離油井半途處，試著整修山路步道。樂協台從彰化來到此地。室內華氏85度；下午5點，室外溫度華氏104度（攝氏40度）。

June 12: Wednesday Went part way to the well fixed the path where we could The colones (colonel) from Changhwa came here today There (Therm.) 85 ° Five P M (P.M.) 104 ° in the sun.

註 氣溫高達攝氏40度，可能筆誤。行文語氣也未強調很熱；但顯然比85度高，是否94度（攝氏34.4度）之誤？是與否，皆為他在台灣時最高溫記載。

6月13日

到油泉上工，雨勢太大，中午下山返家。整個下午與晚上部分時間，老樂都與我們在一起。華氏83度。

June 13: Went up to the oil springs rained so hard came back at noon The colonel stayed by us all the rest of the day & part of the night Ther (Therm.) 83 °.

6月14日

老樂與我們同到油井，我們生火、加足蒸氣開始鑽井，將鑽頭桿軸（shaft）鑽入地底深達14呎，再把鑽井衝管（或套管, driving pipe）深入5呎，〔加上先前已鑽的7呎，〕共達26呎深。

June 14: Went up to the well got up steam crilled (drilled) over the shaft 14 feet drove the shaft 14 feet drove the pipe down 5 ft which makes 26 feet in all The colonel went up with us There (Therm.) 84 °.

註 早期衝管用圓木或板材為材料，一段段連接成，如7月16日記載因缺木材、無法製作衝管而停工；目前則用鑄鐵衝管（cast iron driving-pipe）。

6月15日

整天在家休息。8位「印地安人」從山上下到村裡來，他們剃頭辮髮，是已開化的土著。華氏85度。

June 15: Stayed at home all day There was eight indians (Indians) come down & were civilized had ther (their) heads shaved Ther (therm.) 85°.

註 絡克晚年回憶，官方為了免除土著出草，每月付50銀元「保護費」給附近部落，要後者擔任保全工作，監視、防範其它部落的覬覦。這8位原住民或為收到保護費部落的社人。

6月16日

在家休息一天。收到Flora, Sarah, Dave, S. Powell來信，沒有報紙寄來。華氏85度。

June 16: Stayed at home all day Received letters from Flora Sarah Dave & S Powell no paperers (papers).

6月17日

到油泉工地，再度發動引擎鑽井，一直工作到晚間7點，驟雨突降才不得不停工返家。河水暴漲，我們冒險游渡。華氏83度。

June 17: Went up to the oil springs got up steam & drilled until 7 oclock (o'clock) P M (P.M.) when it rained so hard that we had to quit & come home The river raised so high that we had to swim to get accrosst (across) There (Therm.) 83°.

6月18日

整天下大雨，只好待在家裡。寫信給Sarah及Sophy Powell。華氏78度。

June 18: Rained all day so had to stay at home wrote letter to Sarah & Sophy

Powell There (Therm.) 78°.

6月19日

下午雨勢加大，在家寫信給Florence，還去一個廟會繞了一圈。華氏80度。

June 19: More rain this afternoon stayed at home wrote letters to Florence went to a Chinese joss pigone (pageant) 80°.

6月20日

河水太高，我未上工，在家寫信給Seyfang；簡時則上去油井工作。〔他回來告訴我〕有人生火不慎，把引擎室的屋頂、還有幾具風箱（bellows）燒掉了，肇禍者被鞭打屁股。華氏82度。

June 20: Water so high that I stayed at home Wrote to Seyfang Karns went up to the well burnt the roof off of the engine house apoilt (spoiled) the bellows & had one man bambooed 82°.

6月21日

還是待在家裡，簡時仍照常上工去。華氏89度。

June 21: Stayed at home all day Karns went to the oil springs There 89°.

6月22日

星期六。到工地將衝管鑽進18呎5吋深，〔連上次的26呎〕合計44呎5吋。與先前挖的洞穴合計，已深入地下約70呎深了。華氏84度。

June 22: Saturday Went up to the oil springs put in 18-5 feet of drive pipe making 44-5 in all drilled about 70 feet There (Therm.) 84°.

6月23日

〔星期天，〕在家休息一天，收到9份《前鋒報》（*Heralds*）。天氣十分炎熱、華氏89度。

June 23: Stayed at home all day Recd nine Heralds very warm There (Therm.) 89°.

註 6月15日～23日九天中，小絡以各種理由休息7天，這是倦勤的先兆。

6月24日

上油井打洞，才下鑽10呎深就發生坍塌現象，需要多接幾隻衝管進去。華氏80度。

June 24: Went up to the well drilled about 70 (10) feet the well caves will have to put in more pipe There (Therm.) 80°.

6月25日

再接上18呎7吋的衝管，整隻管長合計63呎。油井四周擠滿看熱鬧的漢人，礙手礙腳的，實在讓我們幾乎無法作業。華氏85度。

June 25: Went up to the well drove 18 ft 7 in (in.) pipe making 63 feet in all There was such a crowd of Chinese around the well that it was almost impossible to do anything There (Therm.) 85°.

6月26日

星期三。整天在油井鑽地，鑽深約76呎（按16呎）。華氏84度。

June 26: Wednesday Went up to the well drilled all ady maid (made) about 76 (16) feet There (Therm.) 84°.

註 手稿看起來像16或76，但他們最後才挖到地底397呎，因此本日深度應是16呎。

6月27日

整天鑽挖油井，剪了一條3呎長的鑽井大繩（bull rope）綁機器，鑽出了鹽水（salt water）與一些瓦斯氣。華氏85度。

June 27: Went up to the well drilled all day cut one foot out of the bull rope struck salt water and some gas There (Therm.) 85°.

6月28日

上油井一看，糟糕，發現地下水湧出、淹滿整個洞穴，趕緊鑽挖，忙了一整天。阿兵哥的眷屬打架，鬧到山上來，士兵無心工作、值勤，我們要那些女人先下山、等我們回家再說。華氏83度。

June 28: Went up to the well found the water running over the top drilled all day The soldiers (soldiers') women got to fighting put the soldiers of (off) the lease & told the women to go home There (Therm.) 83°.

6月29日

今天埋頭苦鑽了一整天。華氏86度。

June 29: Went up to the well drilled all day There (Therm.) 86°.

6月30日

星期天。在家休息，再翻閱故鄉的《提塔斯維爾前鋒報》。華氏88度。

June 30: Sunday Stayed in the house all (day) reading the Titusville Herald There (Therm.) 88°.

7月1日

星期一，上工去。再接上20呎7吋的衝管，加起來全部管長共84呎。

官方再從龜崙嶺（Keeluney）派一位新翻譯來。華氏86度。

July 1: Monday Went up to the well drove 20 feet 7 in more pipe making 84 feet in all The Chinese new interpreter came from Keeluney There (Therm.) 86°.

註 新來譯員可能叫小周，詳7月9日。

7月2日

鑽了一整天，再挖深12呎。士兵從龜崙嶺回來，買回麵粉、雪茄，及其它雜貨。華氏86度。

July 2: Went up to the well drilled all day making about 72 (12) ft soldiers came back from Keeluney bring flours cigars & ? There (Therm.) 86°.

註 據簡時〈遠東函〉稱，「台灣麵粉很少，我們須從香港訂貨，要花上三週的等待時間。麵粉一到，就是胃口大開的時候」。

7月3日

再鑽整天，最後蒸氣鍋爐過熱，把抽水機的蓋子衝上半天空。華氏88度。

July 3: Went up to the well drilled all day burnt the boiler & blowed the paching (packing ?) out of the pump There (Therm.) 88°.

7月4日

星期四，美國獨立紀念日，卻無溫純的啤酒（lager）陪伴，鑽一整天油井，應該對得起本年光輝的7月4日吧？華氏90度。

July 4: Thursday And no lager Went up to the well drilled all day so much for the glorious fourth of July Ther (Therm.) 90°.

7月5日

星期四。到油井把〔抽水、抽沙、抽油用的〕164呎長的抽油套管（casing）插進地底。華氏90度。

July 5: Went up to the well put in the 164 feet of casing There (Therm.) 90°.

註 現在的casing又稱熟鐵抽油管（wrought-iron pipe）。

7月6日

今天鑽了26呎左右。晚上村內有人在街上打架，沒人受傷，只聽到很多聲互相「問候」對方長輩的三字經。華氏90度。

July 6: Went up to the well Drilled about 26 feet Had a street fight this evening no one hurt but a good deal of Chinese talk There (Therm.) 90°.

7月7日

〔星期天，〕派士兵到〔4哩外〕大南灣（Tanawan, 可能是南方4哩處的大湖鄉大南村，原稱大、小南勢）買鳳梨，沒買到，卻帶回一顆西瓜。另一兵去後壠，爲我們一口氣買回了18粒鳳梨。氣溫華氏89度。

July 7: Sent soldiers to Tanawan for pine apples (pineapples) could not get any but he brought a water melon (watermelon) One soldier went to Oulan & bought 78 (18) pine apples for us Thereometer (Thermometer) 89°.

7月8日

星期一。上山到油井工地，發現抽油套管因接縫處鬆動、管內滿是水，將管子朝下釘打3、4吋使它接緊，再開始一整天的作業。華氏90度。

July 8: Monday Went up to the well found the casing full of water drove it down three or four inches & made it tight drilled the rest of the day There

(Therm.) 90 °.

7月9日

雖然下著雨，仍到油井鑽地，忙了一整天。回家聽說有來信，原來是小陳（Chang）、小周（Chow）、布郎、唐道台寫信給簡時的公事〔，空歡喜一場〕。華氏88度。

July 9: Went up to the well drilled all day Karns recd letters from Chang Chow Brown & Taotai There (Therm.) 88 °.

註 日記中的Chen, Cheng, Chang可能都是同一人，姑且稱他小陳。

7月10日

今天鑽出更多的鹽水，油井塌陷十分嚴重，必須放入更多的衝管及抽油套管，我很擔心情況不妙。晚間下起雨，華氏87度。

July 10: Went up to the well drilled struck more salt water the well caves very bad will have to put in more drive pipe & casing & am afraid Rain to night (tonight) Ther (Therm.) 87 °.

7月11日

星期四（原文寫成周二）。用抽油套管把鹽水、泥沙抽出，不慎掉了三節油管在井底。試著旋轉衝管軸柱想栓緊、勾起那掉落的油管，搞了半天徒勞無功。華氏85度。

July 11: Tuesday (Thursday) Went to the well drained the casing and left three peaceses (pieces) in the well tryed to to (tried to) screw on to it but could not Thereometer (Thermometer) 85 °.

註 Thursday寫成Tuesday，或為筆誤，但也顯示心情欠佳。

7月12日

用纜繩（rope）把插在小型卡瓦打撈套筒（slip socket）上的一節抽油套管（casing），垂入油井內，試圖釣起那三節掉在井中的油管。搞了兩小時，纜繩斷裂，連卡瓦打撈筒都一齊掉進井底。華氏82度。

July 12: Went to the well made a casing spear out of our small slip socket put it down the well took hold of the casing jared about two hours and broke the rope socket There (Therm.) 82°.

7月13日

今天改用較大的卡瓦打撈套筒、較粗的纜繩，但仍無法勾起掉落井底的寶貝工具。咱們不信邪，用一根油管（tubing）做成鐵勾（hook），再垂入井中，一陣胡攪亂勾，終於在下午5點勾住那三節油管與小型卡瓦打撈筒。小心翼翼地一吋吋慢慢拉起，深恐不小心又得重來，搞到半夜1點，才算苦功告成。鬆了口氣，用大夾鉗（clamps）夾住油管、小工具，吊在井架處；再把卡瓦打撈筒與衝管掛在牆上。今晚，也可說凌晨3點才爬上床。

July 13: Put the large slip socket in the large tools changed the rope end for end but could not get hold of the tools in the well made a hook out of tubing put it down the well on tubing straightened the tools up then run (ran) the large socket got a hitch 5 oclock (o'clock) p m (pm or p.m.) had to jar it the most of the way up got it to the top of the well at 1 oclock mid night with the small tools & the casing put the clamps on the top of the casing & let it hang in the well with the tools in side One of slips on the large socket got on the out side & enf (?, hung) on the wall & the drive pipe went to bed 3 oclock.

7月14日

星期天。上午9點起床，累歪了，在家休息。今天是農曆〔六月〕十五日，村民做拜拜，燒了很多香、放了很多鞭炮。華氏89度。

July 14: Sunday Got up this morning at 9 oclock (o'clock) stayed in the house all day This is the Chinese 15th of the moon plenty of fire crackers and joss paper There (Therm.) 89°.

註 本日是半年節，大抵為漳州籍移民的祈神祭祖習俗，供物用湯圓、甜糯米飯。公館鄉以客家移民為主，他們可能有些來自漳州客家原鄉，才有此俗。是否？待考。

7月15日

備妥機器，將衝管帽（driving cap）略轉鬆八分之一英寸，使其與衝管密合，再接到引擎飛輪（fly wheel）。接到葉文瀾（Yap Bun Lan）來信，內述他目前人在後壠，明天將上山視察。一隊士兵從彰化抵此。華氏85度。

July 15: Went up to the well took the tools down fasend (fastened) the driving cap to the fly wheel of the engine & turned it down 1/8 of an inch so it would fit the driving cap There (Therm.) 85° Received letter from Yap Bun Lan who is at Oulan & will be here to morrow (tomorrow) Soldiers came from Changhwa.

❶ 葉文瀾曾任廣東候補道、福州造船廠總辦，巡撫丁日昌擬用西法開挖台煤，調葉主持台灣煤務總辦（1876年），兼理石油、硫磺事務。葉與唐景星共同主持台油鑽鑿事宜，但真正主事者為唐，根據络克的記載，葉文瀾僅到此一次，前來略盡職守。相對於煤務，據英國領事貿易報告，葉因經常在廈門常住，經營自己的事業，外務太多，不太過問煤務，可能從下過煤坑視察，英國工程師翟薩（David Tyzack）又未被賦予實權。

後來忍無可忍的翟薩向前來巡視的新任福建巡撫吳贊誠、台灣道台夏獻綸打小報告（吳贊誠任職巡撫期間，曾於1877年6至9月及1878年10至12月間兩度巡視台灣，葉於1878年6月請辭，因此翟薩打小報告可能在1877年那次；但吳那次似未到基隆），葉遂於1878年6月託病請辭。稍後，基隆通判何恩綺（Ho Ngên-ch'i）兼任，卻因基隆地區環境不健康而移駐艋舺；同年12月中起接任者即是新任的基隆通判鄭膺杰（Chêng Ying-Ch'i）。參閱 "Tamsui Trade Report, for the Year 1878," pp. 217-218 or pp. 總346-347；〈1877年英國駐淡水領事與貿易報告〉, p. 144, IUP, China 12, p. 366；〈1878年英國駐淡水領事與貿易報告〉, p. 159, IUP, China 12, p. 721.

❷ 絡克稱彰化來的兵都用soldiers，這隊士兵應是所謂的保鑣；稱駐紮油泉的兵為government soldiers, 偶用soldiers。

7月16日

雖然去到油井上工，但缺乏木材，無法架設鑽井衝管。不到中午，即下山返家吃午飯。華氏90度。

July 16: Went up to the well did not do anything as they have not furnished lumber for us to rig our driving pipe came back to tiffin 90°.

7月17日

將推動鑽井衝管的「大鑽頭」（？, driving male）略作調整。葉文瀾上到油井，試圖說服我們搬到油井工寮，與士兵住在一起，但被我們拒絕了。華氏85度。

July 17: Went up to the well hued out a driving mall (male ?) Yap Bun Lan came up tryed (tried) to have us move into the house with the soldiers but we refused. 85°.

註 他們每天來回至少要走2哩路，有時還要冒險度過湍急的河流；不想搬上去的唯一合理解釋就是「快不想幹了」。

7月18日

瘋狂鑿井，用鑽井衝管挖了52呎又3吋。唐道台抵達油井視察。華氏88度。

July 18: Went up to the well drove 52 ft 3 in (in.) drive pipe Taotai came up to the well 88°.

註 這是自1月17日後，唐景星再次上油井，似乎待到8月1日。

7月19日（未記）

7月20日

星期六。試圖將大廣口瓶塞擠入護管內，但太大、塞不下，只好換上較小的。唐道台今天也來到油井看我們工作。雨中步行返家。華氏86度。

July 20: Saturday Went up to the well put on the big jars but could not get them through the drive pipe so we have to use the small ones Taotai came to the well to day (today) There (Therm.) 86° came home in the rain.

7月21日

〔禮拜日，〕在家休息。晚間微雨，華氏88度。

July 21: Stayed at home all day Light rain this evening There (Therm.) 88°.

7月22日

把油井四周邊緣加大、掘深40呎。兩名駐紮油井工寮的士兵（government soldiers）今天〔因病〕死亡，一位死在油井附近、一位走到4哩外的大南灣突然死亡。華氏86度。

July 22: Went up to the well rimed the hole down 40 feet Two government soldiers died to day (today) one at top side and the other walked to Lanawun four miles & then died. There (Therm.) 86°.

註 這是日記首次記載有人死於瘧疾、飲水不衛生所引起的傷寒、森林熱。據〈1878年淡水海關年報〉，當年風土病橫掃北台灣，洋人染上有嘔吐現象，漢人則無，不加治療會致命。稍後疫情蔓延，9月15日簡時、同月22日絡克相繼生病。

7月23日

在油井除了鑽井打洞之外，還能幹甚麼？幹活一整天。華氏86度。

July 23: Went up to the well drilled all day 86°.

7月24日

道台陪我們上工，要苦力回去吃自己，換上無事可做、到處閒晃的士兵做工。華氏88度。

July 24: Went up to the well drilled put the soldiers at work & discharged the coolies The Taotai was at the well to day (today) There (Therm.) 88°.

7月25日

儘管午後大雨，我們仍冒雨幹活。華氏84度。

July 25: Went to the well drilled all day hard rain in the afternoon 84°.

7月26日

油井坍塌十分嚴重，〔忙著弄出塌陷的泥土，〕今天鑽不到3呎深。明天再用抽油套管抽出地底的泥漿、鹽水。今晚下雨，明天河水料必高漲。華氏85度。

July 26: Went up to the well Caves in so bad that did not make three feet will try to put in casing to morrow (tomorrow) raining to night (tonight) expect high water 85°.

7月27日

雖是星期天，仍然上工，插入256呎長的抽油套管，抽出不少泥漿、鹽水。午後大雨，下了工，溪水暴漲，只好游泳過河返家。華氏87度。

July 27: Sunday Went up to the well put in 256 feet of casing rained in the afternoon had to swim to the creek to get home There (Therm.) 87°.

7月28日

〔星期天，〕官方人士邀我們一齊到Wetankha（加苳坑——Eltarcau？）去看儲存在那裡的空油桶，但我們不願前往；我寧可在家寫信給Flora與Florence。華氏84度。

July 28: Received an invitation to go to Wetankha to look (at) some oil tanks but would not go Stayed at home wrote letter to Flora and one to Florence There (Therm.) 84°.

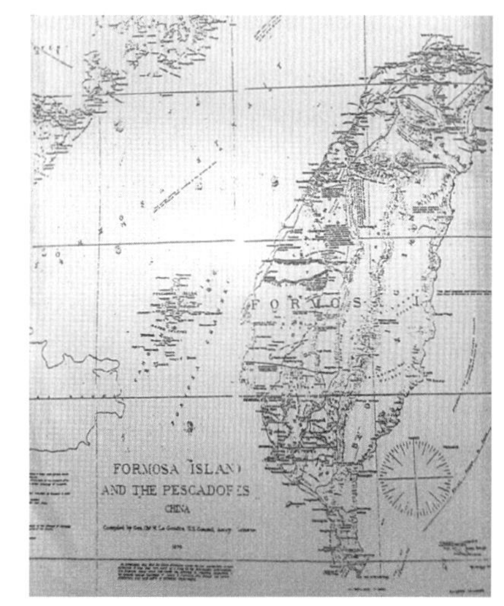

▲李仙得收集之台澎地圖（陳政三翻拍）

❶ 官方認為即將鑿到油，須先準備；老美不願前往，再度露出倦勤徵兆。

❷ Sampson Kuo在論文（頁169、479）認為Wetankha或Aetamkha即是李仙得（C. W. Le Gendre）在地圖標的Altoka，斷定即是原苗栗市舊部落嘉志閣社的

地名稱呼（今苗栗市嘉盛里），該地就在苗栗市。但絡克皆以Marlee或Marley稱苗栗市，而嘉志閣實在與上述英文地名發音差太多，倒是與茄苳坑近似。因此，筆者推測或是4月27日、6月3日提到的茄苳坑。是否？待考。

7月29日

整天忙著鑽井、抽出廢泥漿，進行尚稱順利。整個下午淫雨霏霏，幸好不大，河水水位沒升太高。華氏83度。

July 29: Went up (to) the well drilled all day casing all right so far Rained all afternoon but did not raise the river much There (Therm.) 83°.

7月30日

（因人在油井，未記，次日補寫）

7月31日

星期三。昨天我未賴在住處，像傻瓜似的冒雨上工，傍晚河水暴漲，無法渡河。如果折返油泉工地，那裡又沒食物，只能另謀下榻之地：先向村民買隻雞、幾個甘薯充飢，再爬過一座1,000英尺的山，終於找到某漢人的茅屋借住一宿。今晨河水水位已稍退，與簡時游泳過河、濕淋淋地回到溫暖的家。唐道台今天到住處來看我們。

July 31: Wednesday Yesterday I was not here I went to the well in the rain like a fool & river rose so high that I could not get back there was no rice or provision at the oil spring so could not stay there Bought a chicken & some sweet potatoes for dinner & then climed (climbed) over the top of a mountain 1000 feet high & stayed in a Chinese hut al (all) night The river fell so during the night that we could swim across this morning Taotai came to day (today).

註 他們為何不與士兵同住油井工寮？從這個小地方可看出老美與government soldiers關係不佳。參閱本年5月23日。

8月1日

周四。抽油套管接縫處不斷滲水出來，將管子再打進地底3呎，使它不漏水。午後天公落水，道台來油井視察。返仕處，收到David及Sarah Locke的來信，以及一大堆報紙。遺憾的是，從淡水（Tamsui）返營的士兵沒帶回麵粉。

August 1: Thursday Went up to the well casing was leaking drove it down three feet & made it tight Rained in the afternoon Taotai came to the well Received letters from David & Sarah Locke Received a number of papers Soldiers came back from Tamsui did not bring any flour.

註 吳子光《一肚皮集》記載：「貓內山……起營汛，幕夷人為工師。其法：就該地鑿一井，徑僅尺許；鑄鐵管如囱，每段長丈餘，逐層銜接；用鐵錐重可千斤，旁以木架、繩索為轆轤轉之。令錐下擊，所遇粗沙大石，俱糜脆成泥，真巧思。其井深數十丈，油日所出數百斤」。參閱《苗栗縣志》，頁253。

8月2日

埋頭苦鑽一整天，打到一片厚達15呎的沙層、鑽透它。下工途中傾盆大雨，我們成了落湯雞。華氏82度。

August 2: Went to the well drilled all day struck a sand went through it about 15 feet thick Rained very hard in coming home got wet through There There (Therm.) 82°.

8月3日（1878年）

星期六。打到油層，但流出的鹽水比石油還多。毫無例外的，在雨中漫步回家。

August 3 (1878): Saturday Struck oil to day (today) but more salt water than oil came home in the rain as usual.

❶ 絡克在台灣以鑽鑿方式取得「第一滴油」的日子，特別標上年代。但8月2日～13日都未提及唐景星，後者直到14日才再出現，唐可能已北上基隆，否則本日挖到「第一滴油」時，不可能缺席。

❷ 根據〈1878年淡水海關年報〉, p. 221或總350頁，當時深度為380英尺。

8月4日

〔星期日，〕在家休息一天，下午又是慣見的天公落水。華氏83度。

August 4: Stayed at home all day rained in the afternoon as usual There (Therm.) 83°.

8月5日

油井冒滿鹽水，上面漂浮些許石油，繼續努力鑽井整天。下午雨勢加大。華氏80度。

August 5: Went to the well salt water running over the top with a little oil drilled all day more rain in the afternoon There (Therm.) 80°.

8月6日

繼續鑽井。官方派工人（Chinese）撈取漂浮油井水面的石油，約得20加侖石油，他們認爲已鑽到了源源不絕的致富油礦。氣溫華氏84度。

August 6: Went to the well drilled all day The Chinese diped (dipped) al (all) the oil that they could get about 20 gallons in all thay (they) think they have a bonanza There (Therm.) 84°.

註 此處用Chinese而非soldiers，可能7月24日被遣散的工人又回來了。參閱本年9月14日。

8月7日

仍繼續鑽鑿。每天約可撈得70加侖，他們還擔心油桶、臉盆不夠用來裝石油。華氏86度。

August 7: Went to the well drilled all day The Chinese dip all the oil about 70 gallons per day they are afraid that thay (they) cannot get tubs enough to hold it Ther (Therm.) 86°.

註 絡克晚年回憶表示，他與簡時曾製作幾個大木桶裝油；還說本想建一座煉油廠。建油廠之事似乎在吹噓。

8月8日

鑽不多久，油井塌陷十分嚴重，加上雨勢越下越大，只好暫停。氣溫華氏84度。

August 8: Went to the well The well caved so bad that we did not drill much more rain There (Therm.) 84°.

8月9日

油井快速塌陷，我們整天忙著清理廢土，沒空、也無法鑽井。今天雨停了，華氏86度。

August 9: Went up to the well did not drill any the well caved in so fast that it took all our time to clean it out There (Therm.) 86° no rain to day (today).

8月10日

周六。到油井，繼續像昨天一樣，清理（他用drilled, 昨天未鑽挖，似爲cleaned之誤）塌陷的土方。氣溫86度。

August 10: Saturday Went to the well drilled the same way that we did yesterday Theremometer (Thermometer) 86°.

8月11日

〔星期天，〕在家休息，今年首次刮鬍子，把留了8個半月的鬍鬚刮掉，並擦拭唐道台送我們防身用的雷明頓步槍（Remington rifle）。

August 11: Stayed at home shaved for the first time this year cleaned the Remington rifle that the Taotai gave us.

❶ 自去年12月27日，兩人相互理髮迄今，未曾記載再度剪過頭髮。假如小絡未再剪髮、未刮鬍鬚，又一度穿睡衣上工，那該是什麼畫面？

❷ 他原有一把左輪手槍，詳1月11日。Remington rifle為186至70年代流行的美製後膛裝彈（breechloader）單發來福槍。

8月12日

整天忙著清理廢土，但才清畢又坍方，忙得火冒三丈。收到淡水（Tomsuia）帶來的一布袋麵粉。接獲Flora的來信。華氏87度。

August 12: Went to the well cleaned out all day as fast as we would cleaned out it would cave in Received sack of flour from Tomsuia (Tamsui) There (Therm.) 87° Received letter from Flora.

8月13日

把抽油管（tubing）放入井中，一下子就被泥巴阻塞住，無法抽出水來。華氏84度。

August 13: Puttubing in the well but it cloged (clogged) with mud & would not pump There (Therm.) 84°.

8月14日

清理油管內的汙水，並把廢泥清理乾淨，發動引擎抽水，情況良好，只抽出少量的油，大部分都是爛泥巴。唐道台搬上來，住在油泉工寮。

August 14: Went to the well drained tubing & put it back pumped all right very little oil The Taotai moved to the oil spring.

註 本日唐景星再度出現，可能是第三度上到油井，停留到10月10日。

8月15日

換掉油管螺帽（caps），連續抽了24小時，抽到6大桶量（barrels, 252加侖）的石油。爬到山上，找尋瓦斯從何處冒出，但只在石縫找到噴出量不多的瓦斯。砍了一節老藤（rattan, 原文用Rutan）帶回家，準備攜回美國當紀念品。華氏83度。

August 15: Went to the well Changed the cups (caps or nuts) pumped six barrels in 24 hours went up on the mountain to see where the gas came out of the rock but did not find much cut a peace (piece) of Rutan (rattan) & brought it home with me intended to take it to America There (Therm.) 83°.

8月16日

與昨日做同樣的工作，不停抽油、抽水，當然夾帶著泥沙。從一條新發現的山路返家，須爬過山頂，累歪了，以後再也不走那山路了。華氏83度。

August 16: Went to the well pumping about the same as yesterday came home by a new road over the top of the mountain dont (don't) want to come that way again Ther (Therm.) 83 °.

8月17日

拉出抽油桿（rods），更換新螺帽。毛毛雨，華氏84度。

August 17: Went to the well drawed (drew) the rods and put on new cups (caps) Light rain There (Therm.) 84 °.

8月18日

〔禮拜天，〕在家休息一天。氣溫84度。

August 18: Stayed in the house all day There (Therm.) 84 °.

8月19日

將抽油桿再度拉出、換上新螺帽。其他白天時間與漢人異教徒鬥嘴、賭氣，被他們煩得要死。晚上下雨，華氏83度。

August 19: Went to the well drawed (drew) the rods put on new cups (caps) bugged the heathen Chinese the rest of the day Rain to night (tonight) There (Therm.) 83 °.

8月20日

上工遇到河水暴漲，打道回府。華氏84度。

August 20: Went up the river but did not go to the well on account of high water There (Therm.) 84 °.

8月21日

抽出抽油桿，更換螺絲。晚間微雨，氣溫84度。

August 21: Want (Went) to the well Drawed (drew) rods put on new cups (caps) Light rain this evening There (Therm.) 84°.

8月22日

下工返家，發現駐紮彰化的副將樂文祥（Lock Lie Len 或Locke Tie Yen）又從彰化城來此，而且幹了一件好事——當廚子正煮晚飯時，老樂居然放了一把火，把廚房屋頂給燒了，引起圍觀士兵、村民一陣興奮、聒噪（great excitement）；現在晚上10點，場面已漸平靜，溫度86度。

August 22: Went to the well came home & found Lock Lie Len (Locke Tie Yen ?) Changhwa colonel here while the cook was cooking supper he set the roof of the cook house on fire there was great excitement but it is getting quit (quiet) now 10 P M (P.M.) There (Therm.) 86°.

註 這是首度、也是唯一一次提到老樂的名字，參閱去年12月5日。老樂為了逼他們搬家，使出「火燒紅蓮寺」的手段。

8月23日

上油井旁山坡工寮，打點搬家事宜，派士兵下山把我們的家當搬上來。今晚，終於有了自從抵達大清帝國後，擁有第一個稱得上屬於自己的「家」。好個喬遷之「喜」！華氏84度。

August 23: Went to the well made arrangements to move top side sent the soldiers back for our goods To night (Tonight) we are in our own house for the first time since we have been in China There (Therm.) 84°.

註 他們自1月9日至本日，共住在溪洲庄小廟七個半月。

8月24日

下了整天雨，只好待在工寮。氣溫華氏80度。

August 24: Rained all day Stayed in the house There (Therm.) 80°.

8月25日

〔星期天，〕天氣轉晴。〔經過討論，〕唐道台決定先把抽油套管（casing）從油井拉出來，再放進去，儘可能越深越好。華氏80度。

August 25: To day (Today) is pleasant The Taotai has concluded to pull the casing out of the well & put it deeper if possible There (Therm.) 80°.

8月26日

先把油管（tubing）抽出，但卻無法拉出套管。華氏82度。

August 26: Drawed the tubing and tryed (tried) to draw the casing but it would not come There (Therm.) 82°.

8月27日

順利拉出套管，開始清理油井。華氏82度。

August 27: Draw the casing with out (without) much trouble & commenced to time (rim ?) down There (Therm.) 82°.

8月28日

油井嚴重崩坍，很難把井口擴大（原文rim，似為ream之誤）。晚間，兩名士兵未奉命令、無緣無故亂開槍，每人被罰打屁股40鞭。氣溫華氏84度。

August 28: Well caved very bad could not rim (ream?) much Two soldiers fired ther (their) guns to night (tonight) without orders& were bambooed for it each received 40 strokes There (Therm.) 84°.

8月29日

繼續清土，進度緩慢。華氏83度。

August 29: Worked on well not much head way (headway) There (Therm.) 83°.

8月30日

一再重複同樣清除坍方的工作。12位從茄苳坑（原文Iltankha, 可能是Eltarcau之誤）來的漢族婦女到工地看我們。

August 30: Worked on well same thing over and over Twelve Chinese women came to see us from Iltankha.

註 這些漢家女可能是4月27日他們到茄苳坑認識的，顯然來看熱鬧的。

8月31日

我們清理坍土的速度永遠趕不上油井坍塌的速度。

August 31: Worked on well caves in as fast as we can clean it out.

9月1日

禮拜天，在工寮休息。煙草、雪茄都抽光了，勉強改用漢人抽的煙桿草。華氏84度。

September 1: Sunday Stayed in the house all day tobacco & sagars (cigars) so have to smoke Chinese tobacco There (Therm.) 84°.

9月2日

白忙一場，毫無進展。簡時與廚師的女兒們（cooks girls）之間發生了點「麻煩事」（trouble）。氣溫83度。

September 2: Worked on well make no head way (headway) Karns has trouble with cooks girls There (Therm.) 83°.

註 原文cooks有誤、忘了標所有格，他們有兩位廚子，一位正牌的阿溪、一位客串的阿三（詳2月20日），所以可能是cook's girls或cooks' girls；前者表示與同一位廚師的至少兩個女兒「有麻煩」，後者為與兩名廚子的至少各一位女兒「起糾紛」。但絡克未提及阿三帶眷屬北上，有可能是後來在當地請的阿溪的兩位女兒。雖語意曖昧，但簡時顯然涉及桃色「劈腿事件」。已婚的簡時在〈遠東函〉提到，有次官員曾建議他把太太從美國接來；但他沒有，所以才造成「麻煩事」的發生。

9月3日

今天工作略有進展，也完成與清國官方的一年合約。華氏84度。

September 3: Worked on well made some progress This finished my year with Chinese There (Therm.) 84°.

註 他們的一年期合約，從去年9月4日離開故鄉開始算起。

9月4日

把坍陷的油井井口擴挖約20呎寬。下著毛毛雨，氣溫華氏83度。

September 4: Worked on well rimed (reamed?) about 20 feet to day (today) Light rain There (Therm.) 83°.

9月5日

清理昨天下工後井口又塌陷的泥土，進度緩慢。華氏84度。

September 5: Worked on well cleaning out what it caved in yesterday slow work There (Therm.) 84°.

9月6日

油井內塞滿爛泥巴，把井口四周堵住，以免更多坍塌的泥巴進來。

華氏84度。

September 6: Worked on well dam the well it is full of mud There (Therm.) 84°.

9月7日

雖然拼命工作，但對這口井已無可奈何。收到Sarah, Dave, William的來信，還寄來很多家鄉的報紙。

September 7: Worked on well but made no progress Received letter (letters) from Sarah Dave and William also lot (lots) of papers.

9月8日

星期天，在家休息一天。布郎寫給簡時的信中說，他將在幾天內到油井來。我決定等他一個禮拜，如果屆時他沒來，那我就自行到淡水去。

September 8: Sunday Stayed at home all day Karns read letter from Brown saying that he would be here in a few days have concluded to wate (wait) for him one week if he dont (don't) come then I shall go to Tamsui.

9月9日

雖然打起精神上工，但卻無法改善目前的惡劣情況。收到布郎給我的信，內容也是說幾天內他即將抵達。華氏80度。

September 9: Worked on well no change for the better Received letter from Brown saying that he would be here in a few days 80°.

9月10日

今天油井的情況略有改善，午後下雨。氣溫80度。

September 10: Worked on well prospects are a little better Rained this afternoon There (Therm.) 80°.

註 未特別註明時間的氣溫，概指絡克晚間寫日記當時的室內溫度。

9月11日

繼續搞那口爛井，情況未變。華氏80度。

September 11: Worked on well no change There (Therm.) 80°.

9月12日

油井情形仍是那副「簡時不疼，我不愛」的死樣子。第一號廚子（Cook No. 1）外出買東西，卻整晚未歸。第二號廚子（Cook No. 2）臥病在床，所以我們無飯可吃，只好束緊腰帶。華氏80度。

September 12: Worked on well all the same Cook No (No.)1 went to buy something & cicnt (didn't) come back Cook No 2 is sick in bed no chow chow 80°.

9月13日

繼續與那油井搏鬥。一號廚子終於回來了。油井四周的漢人都染病，平均每天打擺子三次。華氏80度。

September 13: Worked on well No 1 cook came back All the Chinese are sick about three times a day on average There (Therm.) 80°.

9月14日

3位生鮮臉的苦力來幫忙。除了1名士兵外，其他士兵都病倒了，得設法使他們康復起來。華氏87度。

September 14: Worked on well Three new coolies on the well Soldiers all down but one set em (them) up again There (Therm.) 87°.

9月15日

〔星期天，〕下到河邊游泳。〔9月8日迄今已一周，〕與唐道台打賭「布郎是否會來」。簡時生病。華氏82度。

September 15: Went down the river to swim Gambled with the Taotai on Browns coming Karns Kicks (is sick) There (Therm.) 82°.

註 為何要等布郎？因為老美迄今尚未領到任何薪水，他們每人領的每月50美元是生活費。詳本年10月23日、25日。

9月16日

繼續清理油井廢土。〔那位健康的士兵也病了，這下〕所有兵都病倒了。華氏80度。

September 16: Worked on the well all the soldiers sicks (are sick) 80°.

9月17日

將抽油套管一段段銜接好、插入井中，長度較抽出前還長一節。

September 17: Worked on well put the casing in one joint more than we took out.

9月18日

再接上另一節套管鑽入井中。整天都下著雨。華氏77度。

September 18: Worked on well put in one joint of casing Rained all day 77°.

9月19日

放入井中的套管鬆動，又無法將它旋緊，只好先拉出，用工具伸入套管，試圖栓緊管內的活環，直到午夜纜繩斷裂爲止。華氏75度。

September 19: Could not get the casing tight so pulled it out put the tools in got them fast jared on them until midnight when the rope broke 75°.

9月20日

改用3吋半的鑽頭插入套管，試圖用卡瓦打撈套筒（slip socket）勾住昨晚掉在井底工具上的纜繩套筒（rope socket），但昨天掉在井中的工具整個被塌陷的石頭覆蓋住了，終歸白費力氣。華氏80度。

September 20: Strung the 3½ in (ins.=inches) tools but could not get the slip socket over the rope socket on account of rock that had caved in on top of the tools 80°.

9月21日

繼續試圖釣起（fishing）掉進井中的工具，但後者就是不吃餌（could not get a bite）。華氏79度。

September 21: Worked on well fishiong for tools but could not get a bite There (Therm.) 79°.

❶ 由於無法弄出井底工具，這口井遭到廢棄的宿命。這是絡克最後一次上工，據他晚年回憶，鑽入地底397呎；如據當年〈淡水海關年報〉，井深394呎，共抽出約400擔（8,800加侖）的石油，其中100擔供糖廍照明用，其餘300擔儲存於後壠。

❷ 《一肚皮集》載，「未幾，井底鐵管被敲擊逼仰，氣閉塞不復通一竅；水齧石泐，鐵錐中斷，萬夫拔之莫能起。夷人目眙氣結，口噤吟不能出一聲而休焉。後遂無人敢問津者。」

9月22日

〔星期天，〕我生病了，整天都不想講話。

September 22: I am sicke (sick) to day (today) so keep quiet.

9月23日

今天未上工，以後幾天也將休息，直到病情轉好爲止。

September 23: Did not go to work to day (today) & shall not do anything until I feel better.

9月24日

病懨懨的，大部分時間都躺在屋內。

September 24: Stayed in the (house) most of the day sick.

9月25日

仍病臥床上，下定決心去淡水，請唐道台替我僱轎子。華氏80度。

September 25: Stayed in the house all (day) & in bed most of the time made up my mind to go to Tamsui told the Taotai to get me a chair There (Therm.) 80°.

9月26日

收拾好行李，準備明天到淡水。華氏78度。

September 26: Pack my trunk for Tamsui There (Therm.) 78°.

9月27日

清晨6點離開油泉（oil springs），夜宿後壠北方7哩不知名小鎭（按似是苗栗縣竹南鎭中港）。

September 27: Left the oil spring 6 A M (A.M.) went to a place about 7 miles above Oulan dont (don't) know what the name is stoped (stopped) for the night.

註 6月4日以後，絡克大都以「油井」稱鑽油工地；但在不順遂時，卻常再用「油泉」字樣。簡時仍留油井，至10月7日。

9月28日

清晨5點出發，中午到竹塹（Turek Chaern, 新竹市），再走40華里（14.3英里）抵達此地（按似爲楊梅）。

September 28: I started this morning 5 oclock (o'clock) reached Turek Chaern at noon then came to this place about 40 li from Turek Chaern.

註 病中的絡克無心情注意地名；如根據美國博物學家史蒂瑞（Joseph Steere）1873年11月21日的旅行，當天的行程為中港——竹塹——楊梅；因此依據腳程，小絡似乎夜宿楊梅。請參閱陳政三，〈史蒂瑞走訪台灣行程表〉，收於《歷史月刊》200期，頁77。

9月29日

（星期天）昨晚發燒、整夜未闔眼，猛灌開水，足足有一桶量，破曉前即啓程赴淡水，傍晚6點抵達「此地」（here, 按似指大稻埕），心情大好。

September 29: Last night did not sleep any too much fever drunk a pale (pail) full of water got up before day light (daylight) & started for Tamsui arrive hear (here) 6 P M (P.M.) feeling quit (quite) well.

註 如照先前寫法，「此地」應是淡水，Sampson Kuo（頁230～231）即解讀為淡水；但往下看，卻發現不是，絡克與稍後北上的簡時，是在10月26日才由該地搬到下游10哩處的淡水。因此，最可能住在洋商、洋行集中的大稻埕（北市迪化街附近）。

9月30日

昨晚一夜好眠，今天在家靜養。

September 30: Stayed in house all day Slept very well last night.

尾聲

再見台灣．再見美國

曲終人散

絡克在1878年9月25日、26日提到離開油井要到淡水，29日抵達北部某個地方，應是地名難唸，他以「此地」（here）或「那裡」（there）代替；就像溪洲庄，他與簡時都覺得拗口，而用this place（1月9日）稱呼一樣。因此如不細查，可能誤以爲絡克到了北部即直奔淡水。依據絡克的筆調，那個地方顯然在淡水河旁、距離淡水上游10英里處（10月26日），最可能的地點就是洋商、洋行集中的大稻埕（北市迪化街附近）。他在大稻埕住到10月26日，才搬到淡水。

10月1日，小絡下到淡水拜訪舊識馬偕牧師，以及採油老前輩、同時也兼任美國駐淡水副領事的德約翰。他與「台灣烏龍茶之父」德約翰互動頻繁，日記提到曾向德約翰借了三次書，後者還請他吃飯。簡時終於放棄繼續留在台灣工作的念頭，揮別那對與他發生「麻煩事」的姊妹花，於10月9日來到大稻埕與絡克會合。停留該地的生活十分單調，就診、靜養、看書打發了大部分的時間。他們似乎沒到過北台最熱鬧的艋舺（萬華），更別提風光秀麗、兼可治病的北投溫泉區了。

依據合約規定，清國每三個月須發薪一次；不過官方並未遵守，10月25日美國技師終於一次拿到十三個半月的薪水，以及返美旅費。次日搬到淡水，與怡記洋行經理郭業賜（Mr. Goetz）同住在一艘舊船改裝的船屋上，靜待航次。11月5日搭上「台灣輪」（*Taiwan*）離開淡水，途經台灣府、廈門、汕頭，12日抵達香港。清國官方給他們返美旅費應足以購買二等艙位，但爲了省錢，他們買的居然是最低價的統艙船票，搭的仍然是來程的同艘船「北京城號」（*the City of Peking*）。

11月16日揮別香港，一路記載的儘是航行英里數，將近一個月的航行，12月12日船抵舊金山。次日，簡時先行搭上火車直奔故鄉，神秘的他從此消失在茫茫人海，連台灣史冊也無法查到他的名字。絡克則拜訪住在舊金山的舅父，15日再搭火車南下洛杉磯，轉往小鎮聖塔安那（Santa Anna）探望住在該地的雙親。日記就記到該日爲止。小絡後來娶了地方選美皇后Miss Fannie Allison, 經營鑽油業致富，成了賓州與油鄉名人，晚年接受多家報紙採訪，留下許多寶貴的資料，尤以這本足以媲美郁永河《採硫日記》的台灣《出礦坑鑽油日記》爲然。他健康又長壽（1850～1943），高齡93才過世。

1878年（光緒四年）

10月1日

〔從大稻埕〕下到淡水（Went down to Tamsui），一口氣買了皮鞋、餅乾、咖啡、牛奶，還拜訪馬偕牧師（Rev. MacKay）與美國〔兼任副〕領事德約翰（John Dodd）。今天沒接到任何信件。

October 1: Went down to Tamsui bought shoaes (shoes) crackers coffee milk See Mckey & Dodd the U.S. Copnsul no mail.

❶ 德約翰簡介，參閱5月30日：他在淡水、大稻埕、基隆都有設「寶順洋行」（Dodd & Co.），廈門、上海也有分行。寶順淡水辦事處兼為美國副領事館，此時星條旗迎著海風飄在埤（鼻）仔頭（今淡水空軍氣象聯隊）河口海邊。他的好友、也是提拔他當副領事的李仙得在1869年《廈門與台灣領事報告》（*Report on Amoy and the Island of Formosa*）提到，「1868年（按1865），德約翰在後壠東南約20英里的番區、雞籠沙巖山脈發現石油從山麓縫隙流出……當地人用這種油點燈，或作為創傷藥。我已把石油的樣品寄到紐約的博物館了。」德約翰發現石油在1865年，提供樣品給李仙得應在1868年，所以後者因此對年代有誤解。

❷ 本日是馬偕第四次、也是最後一次出現於日記記載。馬偕未在日記記載油匠來訪，僅載南部長老教會的巴克禮牧師（Thomas Barclay）仍在淡水作客。巴克禮在9月29日搭乘海龍號抵達淡水。*Mackay's Diaries*, p. 253；《馬偕日記》，頁350-351。

10月2日

星期三。寫信給人還在油井的簡時，收到A. T. Irvine的來信。

October 2: Wednesday Wrote letter to Karns & received one from A T Irvine.

10月3日

與美國〔副〕領事德約翰先生共進午餐，由於我的廚子（阿三？）生病，所以很高興可以藉機換個口味。今天凌晨發生兩起火災，凌晨1點發生在住處東邊，凌晨3點（原文寫爲P M, 應是A.M.或a.m.之誤）的在西邊。

October 3: Took tiffin with Mr Dodd U S Consular my cook is sick & I was very glad to get something to eat There was two fires this morning one on the east side of me at 1 oclock (o'clock) A M (A.M.) the other on the west side at 3 P M (P.M.).

❶ 吃午餐的地方不一定在淡水，德約翰也常住大稻埕。

❷ 簡時似乎比較欣賞中國美食，〈遠東函〉提到他如何享受台菜，「他們將包含頭、腳在內的整隻雞放進鍋裡燉爛，只要能吃的東西，絕不浪費」，還說返美後將常上館子享受中式料理。

10月4日

以每月8銀元請了一名新廚子，迄今晚爲止我對他挺滿意的。

October 4: Hired cook at $8.00 per month am well pleased with him so far.

10月5日

星期六。到唐道台〔住所花園接待人客〕的茶樓（tea house）繞一圈，僕人說道台生病了，即將從油井回來。收到E. L. Wheeler, Nett, Flora, Sarah來信，另有很多報紙。

October 5: Saturday Went to the Taotai tea house his man told me that the Taotai was sick & was on his way home Recd letter from E L Wheeler Nett Flora & Sarah also lot (lots) of papers.

註 當時大稻埕為茶行集中地，第一版一刷筆者誤解讀為「唐景星在北台經營茶葉生意」；「tea house」似應為接待人客的茶樓。

10月6日

〔禮拜天，〕因雨，未外出。

October 6: Stayed in the house on account of rain.

10月7日

下整天的雨，還是待在住處。

October 7: Rained all day did not go out.

10月8日

去找德約翰先生，並借了幾本書。

October 8: Went to Mr. Dodd & borrowed some books to read.

10月9日

簡時〔7日晨〕從油井北上，今天抵達此地。

October 9: Karns came up from the oil spring.

❶ 據〈1878年淡水海關年報〉，清國在油泉鑽了兩口井，絡克日記只提到第一口油井：他離開後，簡時只多留10個工作天（9月27日晨～10月6日晚，7日北上、行程三天），是絕不可能把機器移到第二鑽井處開挖的。如有，簡時一定會向小絡提起，而後者也會留下記載或在晚年透露給記者。因此在老美都離開後，唐景星似乎找了其他人鑽鑿第二井，這些

人或許是跟美國技師學會三腳貓功夫的學徒工匠，但效果不彰，旋即放棄。1872年入墾出磺坑的隘勇首邱阿玉，在油井荒廢後，曾一度獲得官府特准，每月繳納稅金30元，採撈舊井湧出的石油販售。

❷ 1887年，福建台灣巡撫劉銘傳設煤油局，交由霧峰林家的「目仔少爺」林朝棟負責，共鑽井5口，但出產量少，入不敷出；89年7月與英商范嘉士（Hankard）簽訂授權開採煤礦、煤油合約，終因朝廷不准作罷。1893～94年間，接任的巡撫卲友濂原擬恢復鑽採，商請他的洋顧問、淡水海關稅務司馬士（Hosea B. Morse）代為物色鑽探人員、評估開辦資本、提出開煉石油的辦法，但哈佛大學畢業的馬士估計開辦資本高達銀洋25萬元，建議不如仍以挖煤為重點，遂終止了繼續鑽油的計劃。

❸ 日治時代，出磺坑油礦從1897年起陸續由淺野總一郎、寶田（1903）、南北（1906）等會社經營開採；1921年由合併改組的日本石油株式會社繼之，到二次大戰末期共鑽98座井，石油產量占當時全台總產量的96%。另外，1904年在出磺坑礦山中設苗栗製油所；1907年移到苗栗街上，出磺坑及錦水的石油、天然氣都以油管輸送到該處處理。日治期尚有新竹州錦水油田（苗栗造橋鄉錦水村）、新竹苗栗郡通宵、竹東郡竹東街員崠子（竹東鎮員崠里）、新竹郡湖口庄、台南新營郡牛肉崎（東山鄉水雲村、林安村）及六重溪（白河鎮六溪里）、台南州嘉義郡中埔臣凍子腳（嘉義中埔村東興村）等油田的開發，屬於海軍省的有高雄州旗山郡甲仙庄（高雄甲仙鄉）及新竹竹東郡寶山庄（寶山鄉寶山村）油田。

❹ 1946年6月1日中國石油公司在上海成立，接管日本在台所有油礦產業，陸續在出磺坑附近發現構造高區深部碧靈頁岩、木山層，以及五指山層之砂岩內，皆蘊藏大量油氣，迄今仍開發生產中，是相當重要的天然油氣礦場。出磺坑油井也是全世界仍在生產的最古老油氣田。

10月10日

星期四。大半時間都在住處。唐道台預計明天會抵此。

October 10: Thursday Stayed in the house most of the day Expect the Taotai to morrow (tomorrow).

10月11日

去找領事〔德約翰〕借了本書；道台沒回來。

October 11: Went to the Consul & borrowed book to read Taotai did not come.

10月12日

整天待在住處〔看書〕。道台今晚從油井工地回來了。

October 12: Stayed in the house all day Taotai arrived this evening.

註 依路程，唐景星最慢須在10日早上離開油井；他從8月14日第三次到油井，至10月10日，似乎都未離開，停留近2個月，可謂盡心盡力，很難得的苦幹實幹型人才。雖然這次鑽油失敗，1879年3月29日英國駐淡水領事費里德（A. Frater）在〈1878年英駐淡水領事貿易報告〉寫道，「據稱唐道台不得不自掏腰包賠給公庫3萬銀元（按約23,100美元）」；但後來他再任輪船招商局總辦，開辦中國鐵路公司、啟新洋灰（水泥）製造廠，均極為成功，對清末自強運動的貢獻，比同學容閎有過之而無不及。

10月13日

〔星期天，〕身體不適，在屋內休息。

October 13: Stayed in the house dont (don't) feel very well.

註 去年本日抵上海。

10月14日

找領事〔德約翰〕借了幾本書。

October 14: Went to the Consul borrowed some books to read.

註 據他的寫法，德約翰這幾天似在大稻埕。

10月15日

無法與唐道台達成任何共識，他提出的方式都不能讓我們滿意。其它時間悶在住處。

October 15: Stayed in the house cant (can't) do anything with the Taotai yet about setting up can get no satisfaction one way or the other.

註 唐景星顯然試圖說服他們續約，所能誘其續留台灣的大概只有加薪、接眷屬來台（小絡尚單身）、提供船票回美探親後再來台等。但他們被瘧疾、不良的工作環境、缺材料、缺後勤補給、官方效率欠佳……等因素嚇壞了，早就萌生去意，更何況直到本日從沒領過薪水。

10月16日

星期三。到一家漢人開的洋貨行買了6瓶啤酒、一把水手刀（大折刀）。無法獲得道台接見。收到Dave, Henry, Alice及父親來信。

October 16: Wednesday Went to a Chinese foreign store bought 6 bottles of ale & a jack knife (jackknife or jack-knife) tryed (tried) to get an interview with the Taotai but it was impossible Recd letters from Dave Henry father & Alice.

10月17日

〔身體不適，〕待在住處。道台請一英國醫生〔凌爾〕來看我，還開了一些藥。

October 17: Stayed in the house all day See the english (English) doctor as he came out the Taotai & had (him ?) send me some medicen (medicine).

註 凌爾（B. S. Ringer），有譯成林格，前者似其正式漢名。1873～80年服務於淡水海關，同時兼任偕醫館（MacKay Hospital at Tamsui）及5家洋行的特聘醫生。

10月18日

在屋內靜養。

October 18: Stayed at home.

10月19日

道台拍胸脯保證，答應在兩天內一定讓我們一次領到所有的薪水（get one pay）。醫生前來出診，但未收錢。

October 19: Taotai makes great promises that we will get one pay in two days the doctor called but would not take any money.

註 合約規定，清方每三個月須發薪一次；不過不但未遵守，而且從未發過薪水。

10月20日

在住處休息。

October 20: Stayed in the house.

10月21日

晚間布郎抵達。

October 21: Brown came this evening.

10月22日

布郎與道台密商，道台說他會解決〔薪水的事〕。

October 22: Brown went in & see (saw) the Taotai who says (said) he will (would) settle.

註 美國技師的薪水，依據合約由去年9月4日起算；絡克保留的領款單副本顯示付到今天為止，共十三個半月。

10月23日

布郎去基隆籌錢，好付清欠我們的薪水與返鄉旅費。

October 23: Brown went to Keelung to get money to pay us off.

10月24日

在住處休息。

October 24: Stayed in the house.

10月25日

布郎帶回籌到錢的好消息，給我們一張〔在廈門的〕怡記洋行（Elles & Co.）開出、紐約取款的薪水匯票領單，還有一張香港領款的旅費匯票。

October 25: Brown came back says everything is alright gave us an order on Elles & Co. for our waged or a draft on N Y (NY or N.Y.) & a check on the bank at Hong Kong for our passage home.

註 絡克保留此兩張由夏道台（Ha Tao-T'ai, 夏獻綸）署名的匯票，他個人計領到十三個半月的薪水、美金1,350元；領到旅費：淡水到香港60墨西哥銀元（約54美元）、香港到舊金山300墨元（美金270元）、舊金山到紐約美金180元。合共1,746美元。依此，簡時應領到薪水3,375美元，旅費則與絡克相同，合共領3,879美元。而他們鑽出的8,800加侖石油當時市價880美元，兩者相差不可以道里計。絡克晚年草估，包括購機器3萬美元，加上所有其它花費，共約10萬美金。Sampson Kuo逐項估計，算出清國官方投資了美金65,437元，但包括了本來就要領薪水的官吏、軍官、士兵的俸給。

10月26日

從原住處那裡（there）到下游10英里外的淡水，道台安排我們候船回國前，暫時與怡記洋行的經理郭業賜（Mr. Goetz）同住在一艘舊船改裝的船屋上。

October 26: Came down to Tamsui proper 10 miles from there we have been stoping (stopping) came on board of an old hulk that has been fitited (fitted) for a house for Elles & Cos (Co's) manager Mr Goetz where the Taotai made arrangements for us to stay until the steamer arrived.

❶ 這證明之前他們不住在淡水；最可能的地方是大稻埕。

❷ 馬偕昨天從關渡返抵淡水。本日日記記載「蔥仔生病（Chhang-a sick）。」27日記載他也生病了。他在淡水停留至30日，該日赴五股坑與甘為霖牧師會合，11月1日南下至中壢，然後到竹塹（新竹）再與甘為霖碰面，於竹塹盤旋多日，再偕同牧師娘蔥仔與弟子至中壢、南崁、新港社（苗栗後龍鎮新民里）、紅毛港（新竹縣新豐鄉新豐村）、後壠、桃仔園（桃園）、新莊、洲裡（蘆州）、五股坑、大龍峒（台北市大同區保安宮一帶）、三重埔、艋舺（萬華）、水返腳（汐止）、基隆、煤港（八斗子）、大稻埕，直到12月27日方再返回淡水。兩名油匠則於11月5日搭船揮別淡水，踏上返鄉之路。*Mackay's Diaries*, pp. 255-260;《馬偕日記》，頁353-359。

10月27日

〔星期天，〕船屋維修、設備均良好，與郭業賜相處愉快。上岸散步。

October 27: Find every thing comfortable & in good shape with Mr Goetz Took a walk on shore.

10月28日

早餐前上岸漫步，飯後在船屋休息整天。

October 28: Took a walk on shore before breakfast stayed on bord (board) the rest of the day.

10月29日

上岸閒逛一小時許。

October 29: Went on shore for about an hour.

10月30日

整天窩在船屋。

October 30: Stayed on bord (board) all day.

10月31日

上到一艘大帆船（sailing vessel）參觀，並在岸邊散步。

October 31: Went on borard (board) of (a) sailing vessel & took a walk on shore.

11月1日

海龍號輪船（steamer *Hailoong*）今天抵港，但她只到廈門〔，不續航至香港〕，所以我們只好再等下航次輪船。

November 1: The steamer Hailoony (*Hailoong*) came to day (today) but as she is only going to Amory (Amoy) we will wait for the next steamer.

註 1871年秋季，登記150呎長、277噸的海龍號（*Hailoong*, the sea dragon）小汽船，開始在法樂船長（John Farrow）指揮下，航行大陸與台灣之間，該船屬於得忌利士洋行（the Douglas Lapraik Steamship Company），是第一艘維持較久、定期往來兩岸的交通船。航線從香港、汕頭、經過中間轉接點廈門，航至淡水、府城、打狗；再由打狗直奔廈門，然後汕頭、香港。1872年進出台灣的客貨量，較1869與1870年倍增，此一時期茶葉輸出增長4倍，因此咸認為有定期航線的需要，而且有利可圖，乃組成得忌利士洋行，由不定期改為定期航線。1872年4月30日費里德（A. Frater）撰寫的〈1871年英駐淡水代理副領事暨貿易報告〉（p. 135, IUP, China 10, p. 243）載，「特別為台灣貿易在英格蘭打造的海龍號汽船於去年（按1871）抵東方，將從下月（1872年5月）起每2週來本港（淡水）載運茶葉」。

11月2日

海龍輪滿載茶葉離港。

November 2: The Hailoony (*Hailoong*) left with a full cargo of tea.

註 當時台茶大多先運廈門、再轉銷國外。

11月3日

〔星期天，〕台灣輪（*Taiwan*）今晚抵港。

November 3: The Taiwan came this evening.

註 500噸的台灣輪原名*Leonor*, 原屬馬尼拉Russell Sturgis公司所有，1874年在香港港外遇颱沉沒；英商得忌利士洋行（Douglas Lapraik & Co.）標購該船，打撈後修復，改名*Taiwan*加入航行台灣——大陸間的船隊，但1882年2月又在澎湖沉沒。海龍輪也屬該公司所有。

11月4日

周一。台灣輪忙著卸貨、裝載茶葉，我們則留在船屋上。

November 4: Monday Stayed on boraed (board) Taiwan discharging cargo & taking in tea.

11月5日

登上台灣輪，傍晚6點駛離淡水，先航往台灣府。

November 5: Went on board of the Taiwan Left for Taiwanfoo 6 P M (PM).

❶ 10月26日搬到淡水迄本日，皆未提到德約翰、馬偕。前者可能在大稻埕，後者如10月26日註，正陪著甘為霖牧師南下巡視教區。

❷ 日記未提到曾碰過淡水海關稅務司李華達（Walter Lay），後者撰的〈1878年淡水海關年報〉，雖未寫出兩位美國技師的名字，但涉及鑽油的部分，資料十分詳細、正確，甚至建議將鑽油工作交由海關主持經營。不知彼此是否會過面？或是李華達聽過德約翰、馬偕、布郎的轉述？

❸ 該年淡水關年報載，兩位美國鑽油技師於本日離港返美。

11月6日

下午3點抵達府城〔安平港〕，心情愉快，沒有暈船。

November 6: Arrived Taiwanfoo 3 P M (P.M.) very pleasant did not get sea sick.

11月7日

傍晚5點航離台灣府，海浪洶湧。

November 7: Left Taiwanfoo 5 P M (P.M.) wether (weather) very rough.

註 停泊26小時期間，他們不去看老王或其他官方人士，倒可以理解；但總應該探訪何藍田領事官、德馬太醫生吧？但絡克居然未提，或是未去道別？5、6兩日，甘為霖牧師在淡水，而如果阿三陪他們到北部，此時應該是彼此揮別的時候。*Mackay's Diaries*, p. 256;《馬偕日記》，頁354。

11月8日

海浪橫掃甲板，是我經歷過最糟的一個海上夜晚，暈船了。上午10點抵廈門。

November 8: Worst night I ever see water washed clear over the ship sea sick arrived in Amoy 10 A M (A.M.).

註 從航程來算，搭同一艘船，從淡水到府城21小時；府城航至廈門才17小時。

11月9日

下船上岸辦事。〔到怡記洋行廈門辦事處〕領到一張紐約銀行的匯票1,300美元，一張舊金山銀行兌現的230美元匯票。往見美國駐廈門領事〔恒德森約瑟〕，領事館未代收到要給我們的信。散步廈門島，四處看看。

November 9: Went on shore recd draft on N Y bank $ 1300 & one on San Francisco for $230 went to see the Consul no mail took a walk on the island.

❶ 這些金額與10月25日的不符，他們應是把扣掉旅費的閒錢，打散成兩張匯票，以策旅途上的方便、安全。

❷ 恒德森約瑟（Joseph J. Henderson），約於1873年下半～1878年冬擔任美國駐廈門領事。1874年8月初，他下令逮捕協助日軍發動「牡丹社事件」的

幕後主謀李仙得（C. W. Le Gendre），轟動全世界，也因而名垂青史。

❸ 廈門當時尚是一個島嶼。

11月10日

星期天。與一群歐洲旅客一道下船吃午餐。晚上8點台灣輪航離廈門。

November 10: Sunday A party of Europeans came of (off) borard (board) to tiffin Left Amoy 8 P M (P.M.).

11月11日

早上8點船抵汕頭（Swatow）；下午4點離港。

November 11: Arrived at Swatoy (Swatow) 8 AM (A.M.) Left Swatoy 4 P M (P.M.).

11月12日

上午8點抵達香港，住進Temprance Hall Hotel。登上〔從美國來時搭的那艘〕北京城號（*the City of Peking*），看看船上〔普通艙〕情形。訂製一套西裝。

November 12: Arrived Hong Kong 8 A M (A.M.) went to the Temprance Hall Hotel Went on board the City of Peking to see how it looked Ordered a suite of clothes.

註 他們為了省錢，購買最便宜的通舖艙位。詳本年11月15日。

11月13日

四處觀光香港城，買些小飾品當紀念。

November 13: Took a look over the town bought a few trinkets to remember the place by.

11月14日

逛公園，還跑上台灣輪再留念一番。

November 14: Went out to the public garden Went on board the Taiwan.

註 不知為何又登上該輪？忘了行李？向船上新交的船員朋友道別？還是藉此看*Taiwan*最後一瞥，以茲緬懷？

11月15日

星期五。購買到舊金山最低價統艙船票（steerage ticket），船務代理商寫了一封信，請北京城輪事務長在途中多多關照我們。

November 15: Friday Bought steerage ticket for San Francisco The agent gave me a letter to the purser to help us along.

註 從美國來時買的應是頭等艙（票價372美元）；返程船票費270美元應可買到二等艙，卻為了省錢坐通舖，這可能即是12日要先上北京城號瞧瞧的原因。

11月16日

搭上北京城號，下午3點揮別香港。

November 16: Left Hong Kong 3 P M (P.M.) in the City of Peking.

11月17日

頂著強烈風浪，本日航行150英里。

November 17: Very rough head winds run 150 miles.

註 他使用英里（mile=1,609米），而非海浬（nautical mile=1,852米）。照說，海上計程似應使用海浬。此處照其使用的字解讀。

11月18日

逆風航行757（按可能是157之誤）英里。

November 18: ? head wind distance run 757 (157 ?) mi.. (miles).

11月19日

航行762（162 ?）英里。

November 19: distance run 762 (162 ?).

11月20日

航行276英里。

November 20: distance run 276.

11月21日

航行290英里。

November 21: distance run 290.

註〈1878年淡水海關年報〉記載，主管鑽油的官員（唐景星）今日離台。

11月22日

航行287英里。

November 22: distance run 287.

11月23日

本日航行305英里，下午5點抵橫濱。這裡相當寒冷，趕緊換穿冬季衣物。氣溫華氏55度（約攝氏12.8度）。

November 23: distance run 305.

Arrived at Yokohama 5 P M (P.M.) quit (quite) cool here had to put on winter clothes There (Therm.) 55.

11月24日

星期日。整天留在船上。

November 24: Sunday Stayed on board all day.

註 去年本日抵達台灣府城（今台南市區）。

11月25日

上岸逛逛，買些古董當紀念品。好冷的天氣。

November 25: Went on shore bought a few curioes very cold.

11月26日

傍晚5點半船離橫濱，今天航行50英里。

November 26: Left Yokohama 5:30 P M (P.M.) run 58 miles.

11月27日

今天航行240英里。

November 27: Distance run 240

11月28日

今天航行243英里。

November 28: Distance run 243

註 去年本日在台灣府城正式簽約。

11月29日

今天航行272英里。

November 29: Distance run 272

11月30日

今天航行322英里。

November 30: Distance run 322

12月1日

〔禮拜日。〕今天航行300英里。

December 1: Distance run 300

12月2日

航行370英里。

December 2: Distance run 370

12月3日

橫度東經180度換日線，所以這禮拜有8天、兩個星期二。航行288英里。

December 3: Cross the 180 line so we will have eight days in this week two Tuesdays. Distance run 288

12月3日

還是星期二、還是12月3日。航行200英里。

December 3: Tuesday Distance run 200

12月4日

航行300英里。

December 4: Distance run 300

12月5日

航行302英里。

December 5: Distance run 302

註 上海發行的《北華捷報》（*The North China Herald*）本日刊出後壠油礦之介紹，並云經過化驗，證實該處生產的確為石油。文中認為後壠地方的交通運輸情況，是可以妥為改善的。

12月6日

航行284英里。

December 6: Distance run 284

12月7日

航行305英里。

December 7: Distance run 305

12月8日

〔星期天，〕航行275英里。

December 8: Distance run 275

12月9日

航行290英里。

December 9: Distance run 290

12月10日

航行266英里。

December 10: Distance run 266

註 去年本日首次探勘出磺坑油泉。

12月11日

航行275英里。

December 11: Distance run 275

12月12日

航行155英里。〔11月16日離開香港，前後經過27天的航行，〕凌晨2點30分，北京城號航抵舊金山美格絲碼頭（Meggs Wharf），須等醫生上船檢疫結束才可下船。上了岸，先投宿布魯克林客棧（Brooklin House）。寫信給Dave及Nett。下午到市區南邊找舅舅哈內柏（Hanables）。

December 12: Distance run 155 Arrived at (San Francisco) Meggs Wharf 2:30 this morning have to wait for the doctor to come and inspect the ship to see if there is any sick on board Went ashore in steam launch & put up at the Brooklin House Wrote letter (letters) to Dave & Nett Went down to uncle Hanables in the afternoon.

12月13日

一早，簡時即先行搭火車東行、返鄉。我則拜訪舅父胡烈斯（Hules），他請吃午飯；拜訪David Farnswouth；晚上哈內柏舅舅帶我去戲院看表演。

December 13: Went to uncall (uncle) Hules to tiffin Went & see David

Farnswouth Went to the Theatre with Uncle Hanable in the evening Karns went east this morning.

註 如據去年從故鄉到金山須8天行程計算，簡時約可於本月20日返抵故鄉提塔斯維爾。

12月14日

把身上剩的墨西哥鷹洋全部以1：0.9的兌率換成美金，買了一雙長統靴。下午4點，搭火車到南加州。

December 14: Sold what Mexican dollars I had for 90 cts Took the train for southern Cal (Cal.) 4 P M (P.M.) Bought pair boots.

1878年12月15日

〔禮拜天，〕下午2點抵洛杉磯。等候另班火車兩小時，轉往聖塔安那（Santa Anna），晚上7點抵達。

December 15, 1878: Arrived in Los Angeles 2 P M (P.M.) wated (Waited) 2 hours for train to Santa Anna arrived 7 P M (P.M.).

註 絡克父母住在聖塔安那。此後，神秘的簡時即消失人間，埋入歷史塵堆，連台灣史冊也無法查到他的名字。小絡則娶了地方選美皇后Miss Fannie Allison，經營鑽油業務致富，成了賓州與油鄉名人，晚年接受多家報紙採訪，留下許多寶貴的資料，尤以這本媲美郁永河《採硫日記》的台灣《出磺坑鑽油日記》為然。他健康又長壽（1850～1943），高齡93才過世。

參考書目

中文書目

《中國石油志》。台北：中國石油公司，1976。
《中國石油工業史料影輯》。中油，1981。
《中國石油工業史料影輯——續篇》。中油，1991。
《油花》雜誌。苗栗：中油，1989年1月號。
《石油工業——台灣經濟奇蹟的主角》。台北：中油．中國石油學會，1992。
《五十年來之中國石油公司》。台北：中油，1996。
《上山下海鑽井忙》。中油，1996。
《天涯何處覓油蹤》。中油，1996。
中華文化復興運動推行委員會主編，《中國近代現代史論集第二十九編，近代歷史上的台灣》。台北：1986。
安倍明義，《台灣地名研究》。台北：武陵，1998。
李仙得，《臺灣番事物產與商務》。台北：台銀台灣文叢第46種，1960。
沈茂蔭纂修，《苗栗縣志》，光緒二十年（1894）。臺灣文獻叢刊第一五九種。台北：宗青，1995。
周璽纂輯，《彰化縣志》，道光十年（1830）。臺灣文獻叢刊第一五六種。台北：宗青，1995。
林子候，《臺灣涉外關係史》。台北：作者自印，1978。
林修澈，《賽夏族史篇》。南投：省文獻會，2000。
林藜，《蓬壺擷勝錄》，上、下冊。台北：自立晚報，1972。
Ruthanne L. Mccunn著，金恆煒、張文翊譯，《悲涼之旅》。台北：時報文化，

1979。
胡光麃，《中國現代化的歷程》。台北：傳記文學，1981。
胡光麃，《影響中國現代化的一百洋客》。台北：傳記文學，1992。
馬偕著，周學普譯，《臺灣六記》。台北：台銀經研室，1960。
馬偕著，陳宏文譯，《馬偕博士日記》。台南：人光，1996。
馬偕著，林昌華等譯，《馬偕日記》。台北：玉山社，2012。
高陽監修，《中國歷代名人勝迹大辭典》。台北：旺文社，1992。
許雪姬，《北京的辮子》。台北：自立晚報社，1993。
連橫，《臺灣通史》。台北：幼獅文化，1977。
黃富三、林滿紅、翁佳音主編，英文版《清末臺灣海關歷年資料（1）、（2）》。台北：中研院臺史所，1997年。
黃嘉謨，《美國與臺灣》。台北：中研院近代史研究所，1979。
黃美金，《泰雅語參考語法》。台北：遠流，2000。
陶德著，陳政三譯述，《北台封鎖記——茶商陶德筆下的清法戰爭》。台北：原民文化，2002。
John Dodd原著，陳政三譯著，《泡茶走西仔反：清法戰爭台灣外記》。台北：台灣書房，2007。
Edward House原著，陳政三譯註，《征臺紀事——牡丹社事件始末》。台北：台灣書房，2008。
陳政三、魏吟冰，《異人的足跡：轉角的風華——陶德》。台北：國史館發行；台北：大康出版，2008。
陳政三，《翺翔福爾摩沙——英國外交官郇和晚清臺灣紀行》。台北：台灣書房，2008。
陳正祥，《臺灣地誌》。台北：南天，1993。
《臺灣地名辭典》。台北：南天，1993。
陳培桂纂修，《淡水廳志》，同治十年（1871）。臺灣文獻叢刊第一七二種。台北：宗青，1995。
國立編譯館主編，《礦冶工程名詞》。台北：國立編譯館，1997。
張德水，《台灣政治、種族、地名沿革》。台北：前衛，2002。

張永利，《賽德克語參考語法》。台北：遠流，2000。

葉美利，《賽夏語參考語法》。台北：遠流，2000。

臺灣省文獻會，《重修臺灣省通志．卷三．住民志同冑篇》。南投：省文獻會，1995。

臺灣省文獻會，《重修臺灣省通志．卷三．住民志地名沿革篇》。南投：省文獻會，1995。

臺灣省文獻會，《重修臺灣省通志．卷七．政治志外事篇》。南投：省文獻會，1998。

臺灣省文獻會，《重修台臺灣省通志．卷九．人物志》。南投：省文獻會，1998。

臺灣省文獻會，《苗栗縣鄉土史料》。南投：省文獻會，1999。

臺灣銀行經濟研究室編，《清季臺灣洋務史料》。臺銀臺灣文叢第278種。南投：省文獻會，1997。

C. Imbault-Huart著，黎烈文譯，《臺灣島之歷史與地誌》。南投：臺銀經研室，1958。

蔡鳳雛，《金門地名調查與研究》。金門：金門縣文化局，2011。

盧德嘉纂，《鳳山縣采訪冊》，光緒十八～二十年。臺灣文獻叢刊第七三種。南投：宗青，1995。

薛紹元纂，《臺灣通志》，光緒二十年。臺灣文獻叢刊第一三〇種。南投：宗青，1995。

黃俊銘，《苗栗縣公館鄉出磺坑石油產業文化景觀保存活化調查研究計畫》。苗栗：苗栗縣文化局，2007。

黃俊銘，《苗栗縣公館鄉出磺坑石油產業文化景觀保存活化調查研究計畫》。苗栗：苗栗縣政府國際文化觀光局，2008。

黃玉雨、黃俊銘、劉彥良〈日治時期苗栗出磺坑礦場設施之發展歷程研究〉，《第五屆臺灣總督府檔案學. 術研討會論文集》。南投：國史館臺灣文獻館，2008。

楊敏芝、周道宏〈產業製程及其保存價值之研究——以苗栗出磺坑爲例〉，《中華民國建築學會第二十一屆第一次建築研究成果發表會論文集》，2009。

外文書目

Brown, H.O.

"ICMC Takow Trade Report, 1875". Taipei: 中研院台史所, 1997.

Campbell, William

Formosa Under the Dutch. London: Kegan Paul, 1903

Sketches From Formosa. London, Edinburgh & N.Y.: 1915;Taipei: SMC reprint,1996.

"The Island of Formosa：Its Past and Future", *The Scottish Geographical Magazine, Aug.*, 1896.

Carrington, George W.

Foreigners in Formosa, 1841～1874. San Francisco: Chinese Material Center, 1978.

Clark, J.D.

Formosa, Shanghai, 1896; 台北：成文reprint, 1971.

Colquhoun, A. R.

"The Physical Geography and Trade of Formosa", *The Journal of the Manchester Geographical Society*, Vol. Ⅲ, Nos. Ⅶ-ⅩⅡ, 1887

Davidson, James

The Island of Formosa, Past and Present. Yokohama: 1903.

De Mente, B.L.

Japan Encyclopedia. NTC, 1995.

Dodd, JoŠ

Journal of A Blockaded Resident in North Formosa, During the Franco-Chinese War, 1884～5. H. K.: 1888.

Hancock, William

"Notes on the Physical Geography, Fauna, etc. of Northern Formosa, with Comparison between that District and Hainan and Other Parts of China", ICMC, "Tamsui Trade Report, 1881".

Harrison, H., 編

Natives of Formosa, British Reports of the Taiwan Indigenous People, 1650-1950. 台北：順

益台灣原住民博物館，2001.

Kuo, Sampson Hsiang-chang

Drilling in Taiwan: A Case Study of Two American Technicians' Contribution to Modernization in Late Nineteenth-Century China. Washington, D. C.: Georgetown University, 1981.

Lay, Walter

"ICMC Tamsui Trade Report,1878". 台北：中研院台史所, 1997.

Ishii, Shinji

"The Island of Formosa and Its Primitive Inhabitants", *The China & The Japan Society*, 1916.

MacKay, George L.

From Far Formosa. Edinburgh & London: 1896; Taipei reprint: SMC, 2002.

Mackay's Diaries: Original English Version, 1871-1901（《馬偕日記英文版》）. Taipei: The Relic Committee of the Northern Synd of the Taiwan Prebyterian Church & Aletheia University, 2007.

Maxwell, William

"Tai-Wan-Foo". H.K.: *Hongkong Journal*, c. 1865.

Otness, Harold M.

One Thousand Westerners in Taiwan, to 1945; A Biographical and Bibliographical Dictionary. 台北：中研院台史所, 1999.

Pickering, William

Pioneering in Formosa, London, 1898; Taipei reprint, SMC, 1993.

Steere, Joseph B.著，李壬癸編

Formosa and Its Inhabitants. 台北：中研院台史所，2002.

索引

人名

地名

鑽油設備

船名

其他

國家圖書館出版品預行編目資料

美國油匠在台灣：1877-78年苗栗出磺坑採油紀行/
陳政三著. -- 初版. --臺北市：台灣書房, 2012.06
面；　公分. --（閱讀台灣；8V16）
ISBN 978-986-6318-68-9 (平裝)

1.石油工業　2.工業史　3.臺灣

457.0933　　101006123

閱讀台灣　　8V16

美國油匠在台灣

作　　者　陳政三(246.4)
主　　編　Meichiao
編　　輯　蔡明慧
封面設計　果實文化設計工作室

發 行 人　楊榮川
出 版 者　台灣書房出版有限公司
地　　址　台北市和平東路２段３３９號４樓
電　　話　０２－２７０５５０６６
傳　　真　０２－２７０６６１００
郵政劃撥　１８８１３８９１
網　　址　http://www.wunan.com.tw
電子郵件　tcp@wunan.com.tw
總 經 銷　朝日文化事業有限公司
地　　址　新北市中和區橋安街１５巷１號７樓
電　　話　０２－２２４９７７１４
傳　　真　０２－２２４９８７１５

顧　　問　元貞聯合法律事務所　張澤平律師

出版日期　2012年6月 初版一刷
定　　價　新台幣300元整